Getting to Grips with Science

A Fresh Approach for the Curious

Getting to Grips with Science

A Fresh Approach for the Curious

Andrew Morris

Imperial College Press

Published by

Imperial College Press
57 Shelton Street
Covent Garden
London WC2H 9HE

Distributed by

World Scientific Publishing Co. Pte. Ltd.
5 Toh Tuck Link, Singapore 596224
USA office: 27 Warren Street, Suite 401-402, Hackensack, NJ 07601
UK office: 57 Shelton Street, Covent Garden, London WC2H 9HE

Library of Congress Cataloging-in-Publication Data
Morris, Andrew (Science teacher)
  Getting to grips with science / Andrew Morris.
    pages cm
  Includes bibliographical references and index.
  ISBN 978-1-78326-591-6 (hardcover : alk. paper) -- ISBN 978-1-78326-592-3 (pbk. : alk. paper)
  1. Science--Study and teaching. 2. Discoveries in science. 3. Effective teaching. I. Title.
  Q181.M875 2015
  507.1--dc23

                                        2014044356

British Library Cataloguing-in-Publication Data
A catalogue record for this book is available from the British Library.

Typeset by Stallion Press
Email: enquiries@stallionpress.com

Printed in Singapore by FuIsland Offset Printing (S) Pte Ltd

For my parents John and Susan Morris who breathed the spirit of enquiry into me.

Acknowledgements

The inspiration for this book dates back to earliest childhood. I hold my parents responsible for my excessive enthusiasm for finding out about things. Was it the environment they created that encouraged me to construct things and look at things all day long or was it just the genes they passed on? It's a hot topic for scientific debate today, but either way it was their fault. The condition was exacerbated by two quite brilliant teachers at Bedwell Primary School in Stevenage, Mr Ewart and Mr Thomas. They appeared to let me just get on with it all day long, whether the "it" was playing the recorder, dividing fractions or making marionettes. It wasn't untill I undertook research for a doctorate in molecular biophysics that I once again enjoyed the kind of rich learning environment that they had provided, the workbench, the wet area, the book cupboard — fertile spaces in primary school and science laboratory alike. I have my wonderful doctoral supervisor, the late Sandy Geddes, to thank for showing me how to blend rigorous scientific scepticism with the excitement of discovery.

The inspiration to take forward my ideas about working with adult learners I owe to my good friend John Vorhaus, who introduced me to the Mary Ward Centre in London. The head of department at the time, Gerald Jones, took the risk of running an innovative science course in a humanities department and I am grateful to him for doing so.

For the preparation of this book I am indebted to the editors at Imperial College Press, Jacqueline Downs, Alice Oven and Thomas Stottor, whose advice and encouragement have been all that an author might wish for. Thanks also go to the friends and colleagues who agreed to read an early draft, spotted errors and suggested helpful improvements — Peter Campbell, Charlotte Eatwell, Will Lake, Daisy Minton, Victoria Minton, Tom Morris, Ian Nash, Lin Norman and Geoff Stanton. For his endless patience and encouragement throughout I am grateful to my partner Franco Carta.

But the stars of the book, as you will see inside, are the members of the discussion groups upon which the entire project has depended. By their enthusiasm and commitment they have demonstrated that simple curiosity, once released, is enough to drive an unending enquiry into the workings of the natural world. By fearlessly articulating their life experiences and persistently probing the scientific explanations offered they have not only developed new scientific insights for themselves but have also forced me into an ever deeper understanding of concepts that I once thought I had grasped. To them I am eternally grateful, especially those who have stayed the course the longest: Peggy Aylett, Ann Boston, Jane Brehony, Hilary Davies, Harry Goldstein, Debbie Karp, Carmen and Susan Kearney, Aileen Cook Nayna Kumari, Monica Lanyado, May Luk, Melissa Rosenbaum, Linda Slack, Paul Treuthardt, Penny Wesson, Emily White and Anna Wojtowicz. Their voices ring out loud and clear throughout the pages that follow.

About the Author

Andrew Morris taught science for twenty years in sixth-form, further and adult education. A physics graduate from University College London, he worked in molecular biophysics at Oxford University and completed a doctorate in biophysics at the University of Leeds. Starting as a teacher of physics and maths, he then became deputy head of Islington Sixth Form Centre and subsequently a director of City and Islington College. Nationally he worked as a research manager in a further education agency and directed a programme aimed at linking educational research to policy and practice. In 2002 he set up an experimental scheme at the Mary Ward Centre to encourage people to explore scientific ideas through the questions that arise in everyday life. The discussion groups arising from this are the inspiration for this book.

Contents

Preface

Who is This Book For?

This book is written primarily for the many millions of people who wish they had acquired a better grasp of basic ideas in science. It is based on a fascinating educational experiment in which people come together with a science teacher to pose questions they have always been curious about and to talk freely about science in their own way. The book may also appeal to people involved in education more generally, simply because it is about an unusual way of approaching science. It explores a radically different way of covering topics (the "curriculum") and of positioning the teacher in relation to the learner (the "pedagogy"). Others may find the content interesting simply because it expresses, often in their own words, how ordinary citizens perceive scientific ideas. The chapters may be read in sequence or dipped into as the reader chooses.

What is it About?

The heart of the book is about what it is in science that interests people with little background in the subject. It explores the wellsprings of people's curiosity and follows the way in which their own personal experiences and preconceptions interact with formal scientific concepts — how frost on a car window might trigger off enquiry about the bonding of water molecules; or how advice on a food label might trigger off enquiry about the structure of lipids. The way in which understanding develops gradually is described, as fundamental concepts are encountered time and again in different contexts. The pathways of discussion are illustrated as they range freely over conventional subject boundaries — physics, biology and chemistry — each playing a part in explaining simple aspects of everyday life, such as why a jumper looks red, how an eel finds its spawning grounds, what defines a hormone. But more than this, the book reveals the astonishing breadth of issues that connect with science as it is conventionally understood. Social issues, philosophy, religion, history, music and art are all encountered in the flow of discussion, driven by real-life questions and experiences.

With the primary readership in mind — people wishing to learn more about scientific ideas — the book describes some of the problems associated with the subject. Difficulties recalled by people who got little from science at school are recounted and compared with findings from research about the nature of science education. Later chapters provide ideas about basic patterns of enquiry that transcend the topics on science syllabuses and offer encouragement and practical advice for people interested in exploring scientific ideas for the first time. This is all based on actual experiences of people in discussion groups grappling with unfamiliar scientific concepts.

The Experiment

The ideas developed in this book and the accompanying extracts from actual dialogues derive from a long-running experimental scheme based in central London. Two-hour discussion sessions are held once a month in informal settings with small groups of people who are keen to learn something about science despite having little background in the subject. The scheme is described in detail in the first chapter. In summary, it involves participants bringing questions or observations from their everyday lives to each session in a spontaneous fashion. The members of the group discuss these initially amongst themselves and subsequently with an experienced science tutor. No topic is prescribed in advance and no topic ruled out. Notes of some 200 such sessions form the basis of many of the ideas put forward in this book.

Why is it Needed?

Science education should be playing an important role in the lives of individuals and the welfare of the country; it is unfortunate that it isn't. Science education is central to tackling two broad national problems: the shortage of people with the skills needed for technical careers, and the low levels of public understanding of scientific matters. These are both serious problems, the ramifications of which extend far beyond the relatively small world of the professional scientist or science educator. The shortage of scientifically trained professionals imperils our health services, engineering, construction and manufacturing industries, though we are currently shielded from the full force of this problem by the skills of people migrating from around the world. The problem of scientific literacy more generally, however, cannot be masked in this way. Scientific advances are driving many political and social debates: genetic modification to increase crop yields, carbon reduction to reduce global warming, DNA testing to personalise health treatments. Willingness to engage with basic scientific ideas is an important aspect of the demands on citizens today. In even broader terms, it is simply a great pity that so many people who enjoy learning about and interpreting the world in which they find themselves miss out on the sheer fascination and beauty of scientific ideas.

The Proposition

The imbalance in understanding between the arts and sciences has long been recognised and the growth of two separate cultures regretted. Children of primary school age do not seem to experience the world in such a divided way and show generally positive attitudes to science.[i] More negative attitudes seem to develop at the lower secondary age and in the English education system the development of arts and science identities becomes enshrined when the spread of subjects is narrowed so dramatically at the age of sixteen.

I was fortunate in experiencing excellent primary education myself, in which making puppets and painting scenery one day would be followed by wiring up the electric lights the next. Finding things out and expressing oneself went hand in hand. Later on, as a college science teacher, I spent many years talking with sixteen-year-olds as they made choices about their future direction. The vast majority identified entirely with either the arts or the sciences, few took a mixture; the strong interdependence of subjects like chemistry, physics and maths exacerbated this trend. My belief was that it was the stark contrast between the nature of the curriculum and teaching approaches in the two cultures that shaped students' choices, rather than an intrinsic preference for the humanistic or naturalistic aspects of the world. Interactive teaching approaches and opportunities for self-expression, plain to see in literature, art and the social sciences, were largely missing in the sciences.

Having held out throughout my life against the pressure to identify with either the arts or the sciences, it seems to me regrettable that the power and elegance of ideas in science is effectively denied to those opting for the arts. If the widespread switch away from science at sixteen is actually a consequence of inflexible teaching and unnecessarily rigid subject matter, might it be possible to represent science within education in a different way? It was with this open question in mind that I hit upon the idea of exploring what learning science might look like, were it to be handled in a more person-centred way. The curiosity that so many adults seem to retain about the natural world was my inspiration. What might happen if this was picked up on and taken seriously in an adult education setting?

Answering that question is the story of this book. A scheme based on this thinking was set up in 2002; people signed up to the discussion groups and took up the challenge with extraordinary gusto. Since then their questions have never stopped flowing. As a result of their boundless curiosity an alternative way of engaging with science has emerged, based not on memorising facts or rehearsing procedures, but on exploring fundamental concepts.

The Role of This Book

The purpose of this book is to offer a preliminary impression of what science looks like when it emerges from people's lived experience. The voices of the participants, noted verbatim at the time, provide the lead in this. They express authentic perceptions about the natural world and the manner in which these link to underlying concepts in science.

The chapters that follow are organised to take the reader through the steps of the process experienced in discussion groups. First, anxieties about science are tackled head on through recollecting difficulties experienced with the subject previously. Thereafter, the kinds of question people ask are described and the paths taken in discussion illustrated in a series of examples. The final sections offer support and encouragement to readers who wish to re-engage with science in adulthood. If some choose to do this the book will have fulfilled its purpose: to help ever greater numbers of people appreciate something of the beauty of ideas in science.

Note: there is a website associated with this book (http://gtgwithscience.com) which provides a place for readers to add their suggestion about helpful books, broadcasts, museums or other resources they have come across.

Endnote

i. Osborne, J.F., Simon, S. and Collins, S. (2003). 'Attitudes Towards Science: A Review of the Literature and its Implications.' *International Journal of Science Education*, 25(9), 1049–1079.

Chapter 1

An Alternative Approach

Natural Curiosity Meets the School Curriculum

What a remarkable resurgence of interest in science we have witnessed in recent years! Popular books on evolution, documentaries on cosmology, exhibitions about the brain: all part of a grand revival. Curiosity about how things work and the nature of the universe we inhabit is alive and well, but it hasn't always been so. Talk of the "white heat of technology" may have characterised the 1960s but subsequent decades saw a growing scepticism or even hostility towards the subject. How can we explain this renewed spirit of curiosity? Are scientists finding better ways to communicate? Are we responding to the extraordinary power of modern graphics and animation technology? Or is society waking up to the importance of science in tackling climate change, food shortages, energy security and other problems in the world? Whatever the cause, the recent explosion of opportunities to enjoy science has made life very much easier for the inquisitive layperson.

This rising interest in the subject is a cause for celebration. It demonstrates the strength and persistence of people's interest in the natural world about them. Even though few actually go on to study science at higher levels, interest seems to remain undimmed for many of those who do not. Despite this, opportunities for people to capitalise on their interest, to deepen their understanding of scientific ideas, are fewer than might be expected. After viewing the TV documentary or visiting the museum what does one do to learn something more of the principles and concepts that underlie the subject? It was with this need in mind that I felt the time was right to organise a scheme aimed at people curious about scientific ideas but lacking a background in the subject.

Unfortunately, for too many people, the experience of science at school was not a happy one. This can easily leave them, as adults wishing to re-engage with the subject, with reservations and anxieties about doing so. They may feel embarrassed about their lack of basic understanding and reluctant to expose it. Yet, paradoxically, these are the same adults who are pumping up demand for popular science books, documentaries and exhibitions. Disappointment about science in education is not confined to those who learn, many teachers also feel hemmed in by the form it takes. For them, trying to convey something of the senses of wonder and discovery that characterise actual science is not easy within the confines of a modern examination syllabus.

Experiences of science at school form the substance of this chapter. Many explanations are available for the switch away from science in the teenage years and many initiatives have been mounted to address these reasons — tackling the supply of good teachers, improving the quality of practical work, marketing physics to girls, and so on. My own sense is that there are deeper causes of the problem; the shortage of good science teachers can't ultimately be overcome until more young people are attracted to study science at higher levels. My contention is that it is the nature of what counts as science in education, and the manner in which it is communicated, that are at the root of the problem. Rigid syllabuses using nineteenth-century concepts of knowledge, coupled with equally archaic notions of instruction, send young people running in the direction of English, history, geography and the arts when faced with narrowing choices at sixteen. These competing subjects have seen immense reforms in recent decades in both their curricula and their approaches to teaching and learning. Subjects in the humanities and social sciences have opened themselves up to learners who are engaged through interpreting information, learning about methods and connecting with real-world contexts. As an important review of science in Europe explains:

> To many a young person, the intellectual edifice [of school science] seems profoundly authoritative and authoritarian — particularly when compared to other school subjects.[i]

So it is not just the shortage of excellent science teachers that is the root problem, nor is it simply the growth of anti-science attitudes in the public. It is the very way in which science is conceived as an educational subject and the way in which it is put across that is at fault. In a nutshell, we need to make science as attractive a choice for teenagers as the arts and humanities are. It needs to allow for self-expression, not purely the absorption of historical facts. It needs to emphasise the importance of ideas, not just of calculations and techniques. Above all, expectations of the teacher as an all-knowing expert need to be altered, and space must be allowed for the more provisional and humanistic aspects of real science to be expressed. It was with these points in mind that the experimental scheme was designed, aimed at attracting people to learn about scientific ideas in a radically different way.

Trying a Different Way

A pilot course was set up in an adult education centre, with meetings held once a week over a short period. The challenge was to find an approach that would release people's natural curiosity and engender a safe, participative atmosphere, yet still provide insight into the core of the natural sciences. The first course proved successful and several members suggested continuing the discussions in an informal setting — a quiet wine bar. The adult education course was repeated over several years and continuing groups were set up from them. Through their growth and success, the experience of these groups has demonstrated the intrinsic value of the approach and thrown light on how ordinary people respond to scientific concepts.

The approach involves discussion rather than instruction and emphasises fundamental ideas rather than factual information and procedures. The very first session focuses deliberately on previous experiences of science learning. It aims to release any negative feelings that may persist and to reassure participants that they will be free to express their thoughts without fear of embarrassment or humiliation. This cathartic experience releases people to think about issues that interest them and emboldens them to pose questions in their own way — questions they would have wanted to ask, had an expert been on tap. It turns out that by simply allowing these to be aired and taken seriously people feel they have taken a momentous step forward, accustomed as they are to feeling overawed by scientific knowledge. Almost immediately questions start flowing, drawing on broadcasts, recent news, books and queries arising in everyday life at home, in the family or at work.

With the issues and questions on the table, discussion soon takes off as points of view are picked up and responded to by members of the group. The simple exchange of perceptions and experiences proves motivating and leads in itself to new insights and challenges to ways of thinking, much as in a discussion of a novel or work of art. At this initial stage "teaching" in the conventional sense is deferred — the task is to build confidence and encourage the expression of diverse points of view. It is upon this foundation of free-running exchange that scientific ideas begin to be built.

Eventually, when sharing within the group has run its course, a thirst develops for some kind of scientific explanation from the tutor. In responding to the plethora of questions, contradictions and misunderstandings that burst forth from the initial exchange, the task for the tutor is to spot the underlying scientific principle in play. A direct answer to a direct question may sometimes be needed, but more often the challenge is to shift discussion from the particular case to the general concept. Thus, for example, a question about the health threat from solar particles hitting the Earth becomes the starting point for discussion about the Earth's magnetism and ultimately about the interaction of electric charge with magnetic fields in general.

Of course, no tutor has specialised knowledge of every scientific field but, as the experiment has shown, this is rarely required at this stage in the process. The most important task is to help people frame their enquiries from everyday life in ways that lead into the scientific body of knowledge. Fundamental concepts that bear on their question, such as gravitation, molecules or cell structure, can then be introduced in such a way that connections are made between the immediate point of curiosity and the broader structures of science. This process alone helps invoke scientific ways of thinking and builds confidence that the grand edifice of science may not be as remote and impenetrable as once thought. If discussion on a topic deepens, more specialised knowledge may well be needed. In the absence of examination pressure or a prescribed syllabus, plans are agreed about finding more detailed information in time for the next session. The tutor, or sometimes participants themselves, may search the internet or read up on a topic between sessions. If interest is particularly high, a visit may be organised to a specialist researcher in a laboratory.

Summary

In summary, the focus of this kind of learning is on ideas rather than the facts and formulae that people so often associate with science learning. Laboratories and mathematical proof are no longer central to the process and the tutor is no longer confined to their own specialist subject area. Knowledge is no longer handed down in a take-it-from-me fashion, nor is it built up in a controlled, systematic way. The driving force is the interest of members of the groups in exploring the science behind their observations in everyday life.

At first sight this approach appears to undermine the very foundations of what we have come to know as science — a hard-won body of knowledge, built up through painstaking experimentation and mathematical deduction over the centuries, requiring a carefully regulated process to ensure its safe transmission to each successive generation. In many respects this criticism is indeed true; this kind of learning cannot be said to be "science" as such. In effect what is being described is a different kind of subject, as distinct from professional science as English literature is from journalism. Akin to many subjects in the "arts" it is designed to engage the broad run of people in an intellectual exploration of the big ideas in a field; it is not intended to prepare a minority of them for a particular career. For that important minority much greater specialisation is needed in addition, involving systematic study of a more traditional kind.

This Book

This book draws on the experience of running such informal learning groups. It analyses the hundreds of discussions that have taken place in them over many years, the questions raised and insights gained. It offers insight into what the scientific issues are that interest people. It addresses the many difficulties people experience in getting to grips with science and describes the kind of excitement they experience — the "epiphany moments", as one group calls them — as a fundamental concept falls into place.

The approach has proved successful on the ground — the groups have shown how interest can be sustained and knowledge built up. In the relatively free area of informal adult education it has sketched out the possibilities of a distinct new approach — science as one of the humanities. A picture of how this might look is built up in the following chapters of the book.

Endnote

i   Osborne, J. and Dillon, J. (2008). 'Science education in Europe: critical reflections. A report to the Nuffield Foundation.' Nuffield Foundation, London, p. 22. Available at: http://www.nuffieldfoundation.org/science-education-europe.

Chapter 2

What Captures Our Interest?

What is it that fascinates us about science? What issues draw us in ever greater numbers to radio and TV programmes and books about evolution, stars or brain science? There are many answers to these questions, and none alone gives the whole story. In this chapter we analyse the experience of people who have been deliberately offered the opportunity to follow their natural curiosity, to ask whatever questions they like about science and to follow them through in discussion with a science teacher. The records of these discussions are drawn upon to provide insight into just what it is that gets us going when it comes to science.

Everyday Life

It's no surprise to find that ordinary people engage with scientific issues in many different ways. The most obvious is the need to get answers to some of the everyday questions of life, questions we normally shrug off as just too complicated to pursue ourselves. In our homes, we wonder how electricity from three small holes is able to pass along a cable and spin our washing machine without leaking out; or even how it got there in the first place. Did it have to travel from the substation down the road or from some remote power station? How long did it take to get here? Who told it to go to my house? How do they know how much I need? What would happen if my child's fingers got into the holes? Countless practical questions arise just from our encounters with household electricity. Then there are the more philosophical ones too — what is it actually? A substance, a force, energy? Is it in our bodies — a vital life force? Is there such a thing as "life force"? In a few simple steps we get from idle queries to a fundamental confrontation with the nature of existence!

This example shows how readily simple curiosity develops into a quest for greater understanding. More than this, it illustrates how quickly a line of enquiry can shift from a technical starting point to an encounter with major humanistic issues. It also demonstrates, as we will see as a thread running through this book, how discussions based on real-life questions draw simultaneously on ideas across the various scientific disciplines: the school subjects of biology, chemistry and physics, and the multitudinous more specialised areas such as pharmacology, neurology, astronomy and materials science. We will also see just how easily discussions lead beyond the natural sciences into ideas in religion, sociology, history, economics and all the other humanities, and how richly the domains complement and reinforce one another.

Another aspect of everyday life that regularly sparks off questions about science is the weather — an inexhaustible source of curiosity in the UK. What causes the wind to blow? Why are rain clouds grey? What is meant by temperature: is 30 degrees twice as hot as 15? Such starting points lead quickly to profound thoughts about the atmosphere in which we live, the nature of gases, of pressure, heat and temperature, and the spin of the Earth in the endless void. Again, a stream of simple questions about the world around us draws us straight into encounters with some of the great concepts of science.

In similar ways, the mildest probing of what our television set is doing for us or how our mobile phone links us to friends a thousand miles away leads us to deep thoughts about the nature of invisible waves, of electromagnetic disturbances and the ways in which our eyes and ears create an illusory model of the world around us. Three simple dots twinkling on our TV screens, red, green and blue, compel us to ponder the meaning of colour and our perception of it. What makes yellow, brown and purple on our screens? How does the illusion of movement arise from static images? What is represented in our brains by the colour red? Discussions of this kind not only help us think through the technical basis of everyday phenomena but also awaken deeper curiosity about the nature of human existence. Simple questions sometimes lead us to confront realities more extraordinary than the strangest fiction. For example, one discussion group started with questions about the meaning of 94.2 FM on a radio dial and why there had been interference one stormy night. Sheer amazement followed as the group began to digest the reality of the countless radio signals that are passing at any moment in all directions through the vast emptiness of space, each travelling at 186,000 miles per second, unimpeded by the walls of our houses; and how these are able to activate our very own radio at the same time as the millions of others tuned in to the same wavelength — truly difficult to imagine and to credit.

However, our interests are not confined solely to technical issues. More personal issues also commonly arise. The health and welfare of people around us — our children and ageing parents, our friends and colleagues — also get us going. Queries about the difference between bacteria and viruses or how babies learn to talk, lead rapidly to discussion of biochemical and microbiological concepts. The nature of cells, their walls and internal structures; the origin of words, the nature of phonemes, the development of brain circuits — such fundamental concepts are taken up enthusiastically, driven by their power to illuminate problems encountered in daily life. Further health-related issues — the effects of hormones on mood, the processes of ageing and the onset of Alzheimer's — all impinge directly on the lives of individuals and each plays a significant part in discussions about science as it affects the ordinary person.

The World Around Us

Of course it is not only the domestic world that inspires our curiosity. The wider world has just the same effect. Daily news stories tell us about the restless movements of our planet, Earth. Natural disasters lead us to question the causes of earthquakes and eruptions of volcanoes. We want to know more about the forces shifting the continents and opening up gashes in the ocean floor. We are mystified by unpredictable volcanic outpourings, where rivers flow with soft liquids we once took for solid rock. The physical nature of this planet, sustaining us, our families and civilisations, arouses our interest in an abstract way. At the same time, the plight of the victims of these disasters raises more practical questions. Are there ways to design houses to withstand tremors? Can tsunamis be detected early enough to provide advanced warning? Can aircraft fly safely through volcanic ash? It seems we have an appetite for both of these kinds of enquiry, the practical and the philosophical. And when we do start enquiring, it is remarkable to discover how the countless branches of science and technology seem to cover just about every conceivable question we might ask.

The range of studies and approaches that have developed over the past 200 years is quite staggering — more than any one of us is able to appreciate. The school timetable concept of science as three distinct disciplines — physics, chemistry and biology — simply fails to convey the richness of the spectrum of knowledge areas and their inter-relatedness: behavioural knowledge from psychology and anthropology; philosophical ideas in neuroscience and quantum theory; historical data from paleontology and DNA analysis; practice-based knowledge in the agricultural and forensic sciences … The list is endless. In tackling major world problems, such as poverty, disease and climate change, policymakers and practical workers on the ground need to draw on knowledge across the board from scientific disciplines, technical innovation, and social and economic studies. Similar cross-disciplinary approaches are needed when we, as individuals, want to pursue our interest in scientific ideas. Only rarely do real-world questions find satisfactory answers within a single discipline; even the simple frying of an egg gets you into chemistry, cell science and thermodynamics before you have even considered the role of taste buds and the pleasure circuits of the brain!

Beyond the physical aspects of the world we live in and the lives we lead lay other issues that strike us principally as aesthetic or moral. Many of these merit consideration in scientific as well as humanistic terms. The sheer majesty of the ocean — the incessant rolling and sucking of the waves, the sparkle of the Sun's glancing rays on the dancing water — draw us into contemplation of the forces at work and their immutability. An eye to the heavens leads to musings about the endlessness of time, enormity of cosmic distances, the place of our little planet and the tribulations we experience on it. Or maybe it's the intricacy of an orchid's flower or elegance of a swallow's manoeuvres that captures our attention. Whatever the cause, aesthetic responses prove yet another way to begin our pursuit of scientific ideas — from the migration paths of summer birds we discover magnetism; from the hexagonal patterning of a sunflower we encounter geometry.

Social and Ethical Issues

Perhaps even more significant as triggers for discussion are the social and ethical concerns that the modern world exposes us to. Images and reports from around the world present us with issues of inequality, injustice and inhumanity at every turn. Gender and race inequalities are commonplace, violence and coercion widespread, and polities often corrupt and oppressive. Once again, scientific insights into the nature of such iniquities need to sit alongside political, economic and social considerations. Advances in the psychology of motivation, crowd behaviour and aggression, studies of brain development and hormonal signalling systems, genetic studies of ethnic groups — all are needed to complement the work of the criminal justice systems, public services and political processes that shape our world. Once again a comprehensive view of knowledge is needed that incorporates aspects we have unfortunately come to see as oppositional — the scientific and technological alongside the humanistic and social. Interdisciplinary study is increasingly important in the academic world and so it is also for the understanding that ordinary people seek, as citizens of the world.

Topic Areas

A selection of the issues brought forward in discussion groups illustrates the tremendous range of starting points for scientific enquiry arising from ordinary curiosity.

The brain: structure, function, nerves, circuits, language, emotion
Hormones: types, mechanisms, moods
Radio, TV, telephones: electromagnetic waves, transmission, receivers, cells
Earth: volcanoes, earthquakes, plate tectonics, magnetism
Gravity: movement, planets, the Moon, tides, the cosmos
Cosmology: the Big Bang, expansion, parallel universes, star formation
Bio-molecules: DNA, proteins, enzymes, drugs
Evolution: genetics, inheritance, early man, evolutionary psychology
Relativity: space, time, motion
Light: colour, perception, seeing
Weather: clouds, rain, the sky, rainbows
Social issues: gender, race, economic interest
Philosophical issues: objectivity, truth, reality

Discussion often begins with a specific question based on an obvious scientific theme, but exploration of it may lead in quite unexpected directions. The courses of discussion are as varied and interesting as the starting points. They range freely in and out of conventional scientific topic areas and often jump, logically or serendipitously, to incidents or observations recollected from the lives of group members. Illustrations of these interesting journeys as they move between personal experience and intuition on the one hand, and experimental evidence on the other, are given in later chapters.

Often it is not just the scientific topic in hand that is pursued in discussion but the very nature of science itself. Questions are raised about how we come to know particular things: What is the historical record? Who found it out? How did they do it? Is it the truth or just an approximate model? Are alternative explanations possible? Did vested interests play a part? When ordinary people discuss science in a free and open way the factual and conceptual aspects of the subjects are integrated with the human aspects of discovery and theorisation. It is this freedom to explore simultaneously the substantive concepts of science and their social, historical and philosophical context that makes the approach so satisfying to participants, and so distinct from their previous experiences of science education.

Summary

In conclusion, the starting points for discussion of scientific ideas are, indeed, many and various. There are countless issues that arise in everyday life for which some kind of scientifically inspired mode of thought is appropriate. Unfortunately, such ways of thinking are open to far too few people because for many, science has been experienced as difficult and remote, and trying to learn about it has too often proved humiliating. As a result scientific ideas fail to play their full part in the mix of resources people have to draw upon in interpreting the world.

Science education not only leaves too few people with adequate scientific understanding, it also fails to help the individual link their natural curiosity to important scientific concepts. Issues in the world that arouse our curiosity call for responses from many areas of knowledge — the simple act of seeing a red jumper involves ideas from optics, electromagnetism, anatomy, physiology, biochemistry and neuroscience, to say nothing of art and philosophy. Yet, science as formulated in education places emphasis on the separation of subjects; to do so inevitably removes it further from everyday experience.

The possibility of working back from the everyday world to the realm of scientific ideas is an approach that can change the way people engage with science. It makes getting acquainted with scientific ideas far more attractive, but entails a reversal of the way we conventionally perceive science education. Examples of this process, based on actual discussions, are given throughout the remainder of this book.

Chapter 3

Past Difficulties

"Don't ask me, I was no good at science at school." How often we hear rejoinders of this kind in everyday conversation! Indeed it is so common that many people feel little compunction in admitting it. In this chapter we consider the difficulties expressed by people in discussion groups about their previous experiences of science education. Many factors are involved, including teaching style, syllabus content, examination pressure and gender stereotyping, yet the accounts given over many years by people of differing ages and backgrounds tell a remarkably consistent story. What follows is based on these personal recollections and on some of the more systematic evidence gathered through academic studies of science education.

In my experience people generally perceive the difficulties they experienced as due to an individual weakness of their own, though in reality it is a problem shared by the majority. It is often a matter of regret to otherwise reasonably well-informed adults just how little they feel they know about science. On further probing this is often found to be linked to unhappy memories of science lessons at school and research tends to bear out such anecdotal observations. An important review of research on attitudes to science states:

> Research findings provide strong confirmatory evidence for children's and adults' anecdotal stories about the influence of teachers on students' attitudes to school science and on subject choice. They raise substantial questions about why the pedagogy of some science teachers is so unappealing to the majority of students suggesting that, whilst science teachers may be knowledgeable about their subject, they are failing to achieve their primary task of establishing a range of varied learning opportunities and communicating their subject effectively.[i]

The danger is that people will too easily blame themselves for their shortcomings in science and so fail to re-engage with it as an adult, to avoid opening up old wounds. For this reason it is important to explore the obstacles people experienced in past learning as a preliminary to re-opening the subject. With this in mind the scheme upon which this book is based includes explicit opportunities for participants to share their previous experience of science learning. In this way feelings of doubt and inadequacy or even anger can be discharged and the way paved for a more courageous approach to exploring science as an adult. This helps ensure that the many strange and challenging concepts that are thrown up in enquiring into science can be confronted with confidence.

The overwhelming majority of people in the author's science discussion groups talk negatively about their previous experiences of science learning. In many cases this is completely at odds with their experience and achievement in other subjects. Most feel it has set them back, in some cases caused pain and humiliation that sometimes arouses indignation in retrospect. This experience stands in stark contrast to their obvious intrinsic fascination with the subject, which proved strong enough to motivate them to re-engage with the subject as adults.

As a consequence of such early experiences of science education it is hardly surprising to find adults who lack confidence in dealing with scientific ideas whilst being entirely at ease with equally complex political, social, economic or historical issues. As a result they are easily deterred from pursuing their questions or putting forward their thoughts on scientific issues. Such reticence is in itself sufficient to inhibit learning; feeling confident about expressing one's pre-existing conceptions in any subject is an important first step in preparing to take on new ones. Studies of the learning process reveal that it is not so much like filling up a half-empty tank with new knowledge, but more a process of opening up one's pre-existing notions to challenge and revise them in the light of more persuasive evidence. We all know from our encounters with young children how their notions of what causes a cup to fall off a table or where their new baby sister came from are as vivid for them as they are limited in explaining the workings of the world.

Teaching

When adults recall their memories of science at school, the problem they refer to time and again is the dullness of the teaching. Many recall lessons that were purely didactic, a one-way stream of information coming from the teacher. Learning was largely by rote, with notes written on the board, to be transcribed into exercise books. According to a recent review of science education:[ii]

> Students complained that school science consisted of too much repetition and too much copying and note taking. They felt that they had been frogmarched across the scientific landscape with no time to discuss any of the ideas or their implications.

Educational research in recent years has shown repeatedly that such methods of teaching are less effective than participatory ones that engage the learner actively and offer variety of approach. Fortunately, they are less prevalent in today's schools. However, for today's adult with an interest in science it is important to know that problems experienced years ago in learning science at school may not be due to their own shortcomings. To be open to scientific ideas as an adult, it is worth trying to get over any feelings of shame you may have acquired from negative experiences at school over which you had no control.

An important study from the Nuffield Foundation, which reviews research about science education across many European countries,[iii] gives valuable support to what many individuals feel to have been the case for them. On teaching, the report describes how science teaching often fails to:

> … generate a sense of anticipation that accompanies an unfolding narrative. Rather than beginning with what might be called overarching questions, such as "Why do you look like your parents?" or "What does the universe consist of?" school science begins with foundational knowledge — what a cell consists of, the elements of the Solar System, or the laws of motion — ideas which appear to most children as a miscellany of unrelated facts. The bigger picture only unfolds for those who stay the course to the end. Lacking a vision of the goal, however, the result is akin to being on a journey on a train with blacked-out windows, you know you are going somewhere but only the train driver knows where.

The report goes on to show that pedagogy in science lacks variety and the quality of teaching is less engaging in comparison to other school subjects:

> The assessment system (i.e. testing and final examinations) encourages rote learning rather than mastery learning for understanding. "High stakes" assessment results in a pedagogy where breadth and repetition are emphasized at the expense of depth and variety which has a negative effect on student engagement.

Many adults in the science discussion groups recall feeling intimidated or alienated by the science learning process and chose to drop it as soon as their schools' procedures allowed. A key issue from the learner's point of view is how their experience of science compares with that of other subjects. In literature there are opportunities to express one's own views — indeed it may be particularly encouraged in the interpretation of a poem, for example. In history, interpretation of information from records and development of empathy are valued alongside acquisition of facts. In geography or sociology, students are encouraged to make connections with their own experiences of the world and thrive on the motivation it provides. As a summary of a major UK educational research programme puts it:[iv]

> Evidence … suggests that children are interested in school science, but are less interested in science than in other subjects. This trend is not unique to the UK … Unless school science explicitly engages with the enthusiasms and concerns of the many groupings that make up today's students, it will lose their interest.

Gender

The discussion groups that have inspired this book were set up specifically for people who were curious about scientific ideas but lacked a science background. As it turned out in each of the groups that were set up, women considerably outnumbered men. Indeed, one group that has continued informally for more than twelve years is entirely female. This imbalance of gender is in marked contrast to experience from A-levels, where in 2011 girls accounted for the majority of all examination entries (54%) but only a minority of entries in science subjects (45%). In the case of physics the imbalance was even stronger with girls accounting for only 21% of entries.[v] In physics particularly, content is simply too male-orientated.[vi] In the words of the Nuffield review referred to previously:

> Research would suggest physics content interesting to girls is almost always interesting to boys but the reverse is not necessarily true and, moreover, that the content of interest to girls is "by far underrepresented in the curriculum". These data are also supported by other research which would suggest that girls would be interested in a physics curriculum which had more human-related content.[vii]

The experience of open-ended discussion in the adult science groups is that discussion frequently focuses on aspects of physics and the subject appears to be of interest equally to both sexes. In an informal analysis of fifty all-female discussion sessions the author found that approximately half of the questions posed by participants related to the physical sciences.

Other aspects of gender, in addition to the scientific material itself, may also play an important part in how people engage with science. The process of discussion depends crucially on willingness to open up enquiry in a tentative way, to express uncertainty where appropriate and to look for responses from others. This helps initial thoughts and preconceptions to be expressed and misunderstandings to surface. The approach has the advantage that prior knowledge is freely exposed and progress in understanding is made by members of the group pointing out to each other weaknesses in an argument. Ideas are constructed mutually so that deep thinking starts to happen before scientific concepts are presented by the tutor. In discussion groups, women seem to be particularly comfortable with this approach.

An important review of research for the UK Institute of Physics on girls in the physics classroom[viii] finds that:

> Students benefit from opportunities to discuss their ideas with their peers in order to see the value of different perspectives, refine their ideas and enhance collaboration in learning. Girls in particular report that they value this approach.

The contrasting tendency, to limit discussion by responding to open questions with closed answers or recitation of facts, can inhibit exploration of ideas and sharing of uncertainties. A key study of science education in schools[ix] reflects the importance of this for effective learning:

> Teachers … must set up activities and questions that help students to formulate and express their own ideas and then listen to what students say. A crucial aspect of such dialogue-based teaching is to give students a voice … so that they will feel free to express even half-formed or confused ideas.

Research on gender in science for adults is unfortunately rather sparse. However, the review of research on girls studying physics at school, mentioned above, concludes that: "What evidence there is suggests that boys are more likely than girls to dominate class interactions". It also suggests that strategies that encourage open discussion by reducing the rush to answer questions help reduce gender differences:

> Strategies ... such as "wait time" for question responses and "no hands up" questioning techniques have been shown to aid the creation of a gender inclusive pedagogy for girls and boys.

My own conclusion, given the evidence that girls are equally excited by science as boys in the primary school, and are strongly represented in the biological sciences at secondary school, is that there is no inherent problem about science itself being attractive to females. However, gender problems do arise in the way the subject is represented in education and particularly in the way physics has been constructed as a school subject. For the adult wishing to get to grips with scientific ideas, there is no reason that physics should be any less engaging than other branches of science. The challenge is not so much the scientific content but the opportunity to engage in open-ended sharing of thoughts and uncertainties about scientific ideas in a way that works equally well for both genders.

Subject Matter

The quality of teaching is not the only disappointment people talk about when recalling experiences of science at school; the subject matter itself is often cited as a further problem. Many people trying to grapple with scientific ideas afresh as an adult feel that they have to start again at the most elementary stage of understanding. The lack of a basic foundation places the inquisitive adult in a difficult position, perhaps feeling almost ashamed at their inability to even find the words to talk about basic concepts such as electricity, proteins or the Big Bang. This feeling is easy to understand, given that so many people simply cease trying to absorb what is being explained to them at an early age. Research shows that it is in the early adolescent years that the switch off science tends to begin. According to the Nuffield review referred to above, interest in science at age ten has shown to be high, with no gender difference, but by age fourteen it has declined markedly.[x]

This sense of inadequacy in discussing science often stands in marked contrast to a person's confidence in discussing social, political, economic or religious ideas, for example. This psychological setback may go some way to explaining a number of apparently contrasting effects when discussing scientific issues with adults. One reaction to a topic about which one knows little is to give up entirely, to profess no knowledge and to accord to others, who may appear to know a little more, an exaggerated degree of respect. In a sense this can become an alternative way of switching off, by not taking on the challenge oneself, not trying to think things through for oneself. An alternative reaction can be to defend oneself against feelings of inadequacy by effectively rejecting the scientific point out of hand. It is possible for this defensiveness to lead on to broader criticism of the objective nature of science and the perceived shortcomings of scientists. Anecdotal examples of phenomena seen as beyond science — mystical happenings or unexplained cures — may be introduced in an attempt to bring down to size what appears as the intimidating edifice of science. Occasions on which science has got it wrong or scientists have acted corruptly may be seized upon. Of course there are plenty of sound reasons to explore the limitations of science and to criticise scientists, but feeling intimidated by knowledge isn't one of them!

But in the main, the fault lies so often less with the individual's failure to grasp the basics, but for with the subject matter to which they were exposed. A common reflection is that science topics at school bore no connection with everyday life and experience. Occasional memory flashes remain — Latin words for parts of a flower; the changing colour of a piece of litmus paper — but rarely was a concept ever built up or an insight gained that resonated with life outside.

This curriculum problem is no great surprise given that school science curricula have long been dominated by the requirements of universities in admitting students to science degree courses. Inevitably the interests of the majority of learners have been sacrificed to those of the future science professional. As the Nuffield review[xi] expresses it:

> The content of the science curriculum has largely been framed by scientists who see school science as a preparation for entry into university rather than as an education for all. No other curriculum subject serves such a strong dual mandate.

Some adults in the discussion groups report that they managed to mug up science sufficiently to pass the external examinations but never felt they had the faintest understanding of what the concepts really meant — then or now.

As is shown throughout this book, there are alternative ways. Scientific understanding can not only be linked to everyday life, but can indeed be triggered by it. Though it is a characteristic of school curricula in almost all countries, scientific understanding does not have to be developed in a step-by-step accumulation of facts. The great risk with such curricular structures is that anyone missing out on one vital stage may be seriously disadvantaged in any subsequent stage. So illness, emotional distress, moving house, growing pains or boredom may all too easily lead to a collapse in understanding a science subject. Would it not be possible to start with the world in which we actually live, as students do in art, English, sociology or geography, for example, and work towards an understanding of it? The experience of groups of adults who have tried this is described in subsequent chapters of this book. Alternative starting points that relate to people's lives are set out and the ways in which understanding of scientific fundamentals can be developed are exemplified.

Language

Many adults report another difficulty with science at school and indeed with many of the books they have subsequently tried to read or technical people they have tried talking to: the terminology — words they have never heard before, words that sound intimidating, words designed for social advantage rather than clear communication. But isn't there a similar problem in many walks of life? I have always found talk of securities, bonds and annuities difficult to follow in finance, or words like plaintiff, codicil and conveyancing off-putting in legal affairs. But despite this we seem to come to terms with these words when we need them; conveyancing becomes clearer when you need to buy a house, annuity as you approach retirement. Clearly, when other people use words that are unfamiliar to you it can make you feel undereducated, embarrassed or even inferior. There are times when this appears to be exploited in gamesmanship — I am sure even scrupulous lawyers and doctors must be tempted to use Latin or Greek on occasions to obscure meanings they don't really want to convey. There are no doubt other times when people use technical terminology to create an impression of superior learning. It can also be convenient in some contexts for obscure terminology to create an exclusive feeling for particular groups. Exams in Oxford are called "moderations"; barristers work in "chambers".

To help us in understanding science we need to take a deep breath when we encounter unfamiliar terminology. First, it is helpful to realise that scientists themselves may be equally in the dark when it comes to areas outside their own specialism. An astrophysicist is as unlikely as you to know what an endoplasmic reticulum is. Second, with a cool head it is worth trying to distinguish exactly why the terminology is being used. It could simply be that a specialist is inadvertently using closed jargon when speaking to a member of the public — a case of bad communication, which could be politely pointed out. But it could equally be that someone is trying to appear impressive with their superior knowledge — something to be resisted. On the other hand there could be a genuine need to describe something in very precise terms or even an educative intention to introduce new language — things to be encouraged. Working out the reason enables you to decide whether to just move on, ignoring the inappropriate use of language or to make an effort to look up a word or ask for clarification. Some suggestions about how to avoid feeling intimidated by the language of science are explored in Chapter 7. Who knows, in the end you may even find it enjoyable to expand your vocabulary in former no-go areas!

Mathematics

A more fundamental obstacle is the widespread use of equations to explain scientific ideas, given that so few people are confident with algebra. Traditionally, maths is seen as a vital adjunct to learning in the physical sciences (physics and chemistry) and engineering, and very important in the life sciences (biology, physiology, etc.) Because of its elegance and precision it serves important purposes in proving theories, such as gravitation, and in clarifying concepts, such as genetic inheritance. However, the overwhelming majority of the population does not enjoy a level of mathematical understanding that enables them to understand scientific ideas by this means.

So how do we deal with this impasse? Sadly, in the main this majority is effectively excluded in formal science education. The focus is on the minority who can hack the maths at the right age and, more often than not, plan to move into science-based careers. My belief is that this need not be the case — we shouldn't give up on the majority. We must find a way in which some of the riches of scientific ideas can be shared with those without advanced levels of mathematics. The success of the discussion groups demonstrates the possibility of a different kind of science for the majority — based on ideas, observation and demonstration rather than mathematical deduction (or for that matter laboratory experimentation).

The experience of running the discussion groups has shown me, a former sixth-form physics teacher, the extent to which formal instruction relies on mathematical deduction as a proxy for conceptual understanding. In principle this appears to be sound — a good discipline for those who can manage the maths. In attempting to respond to people's simple questions about basic issues of everyday life, however, I see how often in my own education the process of mathematical deduction acted as a diversion from conceptual understanding at a deeper level. In attempting to explain things in a purely conceptual way — for example, why the tides occur twice per day — I can see that the mathematical explanations I learned for exam purposes yielded too superficial an understanding. Of course, some advanced concepts can only be conveyed thoroughly in mathematical terms, but in the main an imaginative as well as mathematical grasp is needed.

For people interested in scientific ideas whose mathematics is at a very basic level, it is important to remain confident that, despite this, there is much to be gained from exploring science. One way to achieve this is simply to ignore people who insist that mathematical explanation is the only way to understand phenomena! There are admittedly some topic areas which are barely accessible without maths, but many others aren't. Books, exhibitions, museums and festivals can be sought out that focus deliberately on ideas and demonstrations rather than mathematical explanation. Popular books about science need to be scrutinised before purchase to check their explanations don't depend on equations that will leave you baffled. In live settings, such as public lectures, exhibitions or museum demonstrations, scientists are increasingly making efforts to explain things using metaphors, images and vocabulary from everyday life. Finally, of course, it is possible that difficulties with algebra or any other area of maths might be tackled head on, perhaps by picking up an introductory book on some fascinating aspect of maths in everyday life, such as probability and chance or the geometry of the coastline. The wonderful world of mathematics might begin to open up as well!

Relevance

It is not only the dullness of teaching and incomprehensibility of the language that so many adults recall from their past experience of science, it is also the lack of relevance to their lives or interests at the time. Tracing rays through imaginary lenses or naming the parts of a protozoon seemed inherently remote things to do as a teenager, however valuable they may come to seem later in life. The Nuffield review cites lack of perceived relevance as a key factor in young people's failure to engage with science:[xii]

> School science is often presented as a set of stepping-stones across the scientific landscape and lacks sufficient exemplars that illustrate the application of science to the contemporary world that surrounds the young person.

But, as experience with both young people and adults alike shows, linking learning to real-life contexts can be very motivating. This is hardly surprising to anyone who has seen the inspirational effect of history brought to life through actual places and records or Shakespeare through interpretation in contemporary settings. Educational research shows it can be true for science too. A systematic review of educational research on approaches to school science that are based on real-life contexts[xiii] showed students' attitudes to science improved without loss of understanding.

A common way to introduce "real-life context" in science education is for the teacher to think of some kind of everyday application that illustrates *post facto* a scientific concept just taught — the thermometer to illustrate the expansion of liquids with temperature, for example. This goes some way to bring science to life but can seem contrived — a kind of sugar-coating to aid the swallowing of a bitter science pill. The alternative approach used in discussion groups *starts* from the real-life question or context the members bring. As we saw in Chapter 2 these may derive from many different aspects of life including philosophical, religious or social issues as well as everyday life or the natural environment. Examples are given in later chapters.

Examinations

The final unhappy memory of school science frequently recalled in discussion groups is the role played by examinations. Clearly for many, the examination simply represented the final humiliation; the proof that you were really cut out for the arts. This unhappy experience for the individuals who then go on to drop science altogether is matched by an equally sad outcome for science and for society at large: a population ending their schooldays with negative impressions of science and painful memories of engaging with it. Society benefits from an adult population that has some basic understanding of the world they inhabit, in science as much as in politics, history or literature. It is equally unfortunate for the science-based professions that so many young people with capabilities of all kinds turn away from scientific or technological careers. Many with the intellectual potential are simply put off by the manner in which the subjects were expressed in school. The widespread and growing enthusiasm for science outside the school curriculum — on TV, in books and through museums — is testament to this. An Ipsos Mori survey in 2011[xiv] showed that the number of people agreeing that "it's important to know about science in my daily life" grew by eight percentage points from 59% to 67% between 2000 and 2011 — a laudable 14% growth.

Unfortunately, the effect of written end-of-year examinations, that were so common when many of today's adults were at school, is to encourage surface rather than deep learning. Teachers are tempted to "teach to the test" by avoiding interesting threads of discussions that are off-syllabus and students are advised to memorise and learn by rote in order to reproduce information in the exam rather than to engage in the struggle to understand at a deeper level. In the discussion groups a few remember actually passing exams in biology and perhaps chemistry, but are adamant that they had no real understanding of the deeper concepts they were regurgitating on paper at the time, and certainly feel they were left with nothing to go on as a conceptual basis for science in adult life.

Summary

This review of common experiences of science at school has been deliberately gloomy. It is based on repeated discussions of memories of science at school with groups of adults wishing to start afresh with science. Time and again the same issues arise about the curriculum, the teaching methods and the effect of examinations. Much of this anecdotal experience is reflected in studies that have been made of science at school.[xv] I have found that encouraging discussion of these negative experiences has been a necessary and beneficial preliminary to engaging afresh as an adult. It has a purgative effect, relieving adults of the obstructive model of science that schooling often bequeaths, leaving them open to new ways of considering the beauty and elegance, the power and the majesty of scientific ideas. From this beginning it is possible for people to grasp fundamental concepts and ways of seeing that match those acquired through literature, social studies and the humanities in general. The remainder of this book is designed to help in this noble and exciting quest.

Endnotes

i.   Osborne, J.F., Simon, S. and Collins, S. (2003). 'Attitudes Towards Science: A Review of the Literature and its Implications.' *International Journal of Science Education*, 25(9), 1049–1079.

ii. TLRP (2011). *Science Education in Schools: Issues, Evidence and Proposals. A Commentary by the Teaching and Learning Research Programme*, p. 7. Available at: www.tlrp.org/pub/documents/TLRP_Science_Commentary_FINAL.pdf.

iii. Osborne, J. and Dillon, J. (2008). 'Science education in Europe: critical reflections. A report to the Nuffield Foundation.' Nuffield Foundation, London, p. 15. Available at: http://www.nuffieldfoundation.org/science-education-europe.

iv. *Science Education in Schools: Issues, Evidence and Proposals. A Commentary by the Teaching and Learning Research Programme*, p. 6. Available at: www.tlrp.org/pub/documents/TLRP_Science_Commentary_FINAL.pdf.

v.   See government statistics at: https://www.gov.uk/government/publications/provisional-gce-or-applied-gce-a-and-as-and-equivalent-examination-results-in-england-academic-year-2010-to-2011.

vi. Murphy, P. and Whitelegg, E. (2006). *Girls in the Physics Classroom: A Review of Research of Participation of Girls in Physics*. Institute of Physics, London. Cited in Osborne, J. F., Simon, S. and Collins, S. (2003) (see endnote i), p. 16.

vii. Osborne, J. and Dillon, J. (2008). 'Science education in Europe: critical reflections. A report to the Nuffield Foundation.' Nuffield Foundation, London, p. 16. Available at: http://www.nuffieldfoundation.org/science-education-europe.

viii. Murphy, P. and Whitelegg, E. (2006). *Girls in the Physics Classroom: A Review of Research of Participation of Girls in Physics*. Institute of Physics, London. Available at: http://www.iop.org/publications/iop/archive/page_41614.html.

ix. TLRP (2011). *Science Education in Schools: Issues, Evidence and Proposals. A Commentary by the Teaching and Learning Research Programme*, p. 7. Available at: www.tlrp.org/pub/documents/TLRP_Science_Commentary_FINAL.pdf.

   x. Osborne, J. and Dillon, J. (2008). 'Science education in Europe: critical reflections. A report to the Nuffield Foundation.' Nuffield Foundation, London, p. 18. Available at: http://www.nuffieldfoundation.org/science-education-europe.

  xi. Osborne, J. and Dillon, J. (2008). 'Science education in Europe: critical reflections. A report to the Nuffield Foundation.' Nuffield Foundation, London, p. 21. Available at: http://www.nuffieldfoundation.org/science-education-europe.

 xii. Osborne, J. and Dillon, J. (2008). 'Science education in Europe: critical reflections. A report to the Nuffield Foundation.' Nuffield Foundation, London, p. 15. Available at: http://www.nuffieldfoundation.org/science-education-europe.

xiii. Bennett, J., Lubben, F. and Hogarth, S. (2007). 'Bring Science to Life: A Synthesis of the Research Evidence on the Effects of Context-Based and STS Approaches to Science Teaching.' *Science Education*, 91(3), 347.

xiv. Ipsos MORI (2011). *Public Attitudes to Science 2011.* Available at: http://www.ipsos-mori.com/researchpublications/researcharchive/2764/Public-attitudes-to-science-2011.aspx.

 xv. Osborne, J. and Dillon, J. (2008). 'Science education in Europe: critical reflections. A report to the Nuffield Foundation.' Nuffield Foundation, London, p. 20. Available at: http://www.nuffieldfoundation.org/science-education-europe.

Chapter 4

Looking at Science Afresh

We have now considered what makes people want to re-engage with science in adult life and looked at some of the difficulties they may have experienced with the subject in the past. In this chapter we look more positively at how perceptions of the subject can be revised, as a preliminary to re-engaging with it in a fresh way. We begin by thinking about how science can seem for people who have not been much involved with it, then move on to how these perceptions might be revised. We take the point of view of a person who is curious about the natural world and wishes to learn, in an unpressurised sort of way, about the fundamental ideas that science has to offer.

The Way It Seems

It's hard

How does the great edifice of science appear to someone who has barely considered the subject since school? Perhaps the most widely held perception is that it's hard. A study by a professor of science education[i] offers several reasons for this perception, among them that: science is abstract, that it involves "reconstructions of meaning" and that there is "insufficient pay-off for the effort involved". A report for the prestigious scientific body, the Royal Society[ii], confirms what many believe intuitively:

> Across all methods science and mathematics subjects were found to be the more difficult compared to non-science subjects. Taking the average difficulty from the different approaches, Coe *et al.* (2008)[iii] found A-level general studies to be the most difficult, followed by physics, chemistry and biology. Overall, it would seem that there is evidence to support the perception of some students that the sciences and mathematics are hard subjects. However, there are clearly many other factors at play in subject choice decisions.

When it comes to individuals talking in a discussion group it is impossible (and probably worthless) trying to disentangle whether their sense that science subjects were harder than others was justified or whether the subjects were just more poorly taught or badly framed. In practice, discussion groups tackle issues that are perceived as hard just as often as ones considered easy. For example, people are as keen to learn about Einstein's theories of relativity and multiple universes as they are about melting ice and floating logs. In the end almost any topic ultimately presents serious intellectual challenges — melting ice leads into the way molecules behave which leads in turn into deeper thoughts about the meaning of temperature and energy. Thus a simple query about a glass of gin and tonic quickly leads to confrontation with profound, abstract concepts.

So yes, scientific explanation rapidly becomes theoretical, however practical the original issue; but is it really unique in this regard? Don't the challenges in understanding the rise of monarchy or the characterisation of Mr D'Arcy also call for complex abstract thought? In coping with the communication of theoretical ideas, experience in discussion groups suggests that it is not so often an inherent difficulty in a concept that hinders progress, but the communicator's tendency to compress too many intellectual steps into one explanation. Concepts are more readily grasped when the steps are broken down and taken slowly, one at a time, with ample use of metaphor to link the new abstractions to familiar images. Intellectual challenges that might otherwise have confused and deterred a learner can in this way become exciting and serve as a stimulus to further enquiry. It has been remarkable to find in discussion groups that when a new concept is eventually grasped there is barely a second's pause to enjoy the moment before the renewed appetite for learning urges further, deeper questioning.

The language is impossible

As touched on in the previous chapter the use of obscure technical language adds to the perception that science is hard. This is a real problem, even bedevilling scientists when they work outside their own field. The alienating effect distances people from materials that might otherwise be relatively easy to understand. However, it is common in many specialist fields — the law, medicine and economics, for example — and to some extent reflects the need for professionals to communicate with precision. The frequent use of Latin and Greek words in science reflects both the origins of science in classical antiquity and also the development of foundation concepts during the eighteenth and nineteenth centuries when higher learning was associated with classical languages. The naming system for species, for example, uses Latin terminology and its creator even adopted a Latinised name, Linnaeus, as was the fashion at the time. Although this may seem burdensome to a modern speaker of the English language, it does reflects the enormous importance of internationalism in science and the value of a universal language in supporting this. Thus, although the English term "Wise Human" might have been an easier term than *Homo sapiens* for speakers of English, the profusion of equivalent names in other languages would surely have led to greater confusion. The opportunity to learn a little classical vocabulary can be seen as a valuable spin-off from decoding scientific terminology. Many complex words are in fact composed of simple Latin of Greek roots. Thus learning that *calor* means heat in Latin helps with understanding words like calorie, calorific or calorimeter, and even the Italian *caldo* or Spanish *caliente*, meaning hot.

It is not only these classical roots that make scientific terminology appear intimidating, excessively long words can have the same effect. But by breaking them down into their component parts they become easier to understand. Knowing just a few root meanings can open up many areas of knowledge. Thus a word such as "leukocyte" can be broken down into *leuko-* (Latin for "white") and *-cyte* (meaning "cell", from the Greek for "hollow container"). This in turn helps in breaking down the meaning of further complex words such as leukaemia or cytoplasm.

It's about three separate subjects

The division of science into discrete subject disciplines presents difficulties when approaching the subject from issues in everyday life. When you want to know more about the effect of cholesterol on the heart, are we talking about the chemistry of a substance, the anatomy of the heart, the physics of blood flow or the nutritional science of fatty foods? As a person with a query we are potentially interested in all of these aspects. The division into subject areas matters less to us as citizens than it does to our teachers. To find out what cholesterol is, what it does and how we might go about reducing our intake, we need to engage freely with the chemistry, physics, anatomy and nutritional science. Unfortunately, this is not how it comes across to us in formal education. The vast empires of knowledge are divided up into broad subject areas of chemistry, biology and physics, which are then divided in turn into branches such as organic chemistry, zoology and astrophysics. Within universities these are divided further into narrower specialisms such as evolutionary genetics, metallurgy or crystallography.

It was not always thus. Up to the eighteenth-century Enlightenment period, which saw huge progress in the sciences, the word "science" was not in common use. All these fields of enquiry were classified as "natural philosophy": the love of knowledge about nature (from Greek *philo* [loving] and *sophia* [knowledge]). It was not until the nineteenth century that the subjects began to separate out, partly because the growing amount of knowledge required greater specialisation and partly because of the external demands of industry and agriculture. The word scientist was invented in the 1840s and the field of physics began to feature as a distinct subject in school examinations in the 1920s.[iv] However, in order to make useful additions to the already mountainous knowledge base a scientist today has to focus ever more narrowly on a specific topic area. For this reason it seems inevitable that the professionals who produce scientific knowledge will work in ever more narrowly defined subject specialisms. For the curious layperson, however, the reverse is probably true. As more and more is understood about how the body works, what determines the climate and how insects stick upside down on ceilings, our question will range ever more freely across discipline boundaries. We want to know what substances are made of at the fundamental level, what they can do, how they are made, how they affect our bodies, where the elements came from and whither we are all destined in the long run! In short, we need natural philosophy!

Interestingly, the sense of science as a group of three subjects is most strongly felt by younger people. An Ipsos MORI survey of attitudes to science in 2011 showed that "younger age groups tend to think [of science] more as biology/chemistry/physics, which perhaps reflects their more recent experience of science at school". Older people, on the other hand, are more likely to associate science with issues such as health, drugs and cures for diseases. So like it or not, the strict allocation of topics into subject areas at school seems more like a device for timetablers and exam designers, than an aid for people wishing to learn about the workings of the natural world.

The order of topics is rigid

A member of one discussion group put her finger on a key feature of science learning that can prove frustrating even for highly motivated students:

> The perception gained, probably from school, is that you have to learn things in order, you have to know one thing before you can know (or even talk about) the next thing. Possibly it stems from the business of laws and to a lesser extent theories. It's as if you have to grasp the law of this or the theory of that before you can discuss the thing that interests you.

As this observation suggests the inflexibility of a sequence of learning can easily lead people to lose interest as they see how far away the topic is that might engage them. To defer issues of interest, perhaps for years, while the rudiments are worked through systematically can leave people feeling disaffected with the subject. A professor of science education once pointed out that he had had to wait till his master's degree to discover that all the atoms of his body had been forged in the interior of stars — a fact that would have been highly motivating had he heard about it ten years earlier as a schoolboy.

Additional barriers can build up over time if a student misses out some step in an interlinked sequence through illness or distraction. Excellent teachers will try to help individuals overcome these kinds of setback, but it is common for adults to describe how they "lost the plot" when they failed to grasp an early point in a sequence that depended upon it. Of course, many aspects of science do require some degree of systematic build-up, in which one step has to be explained before another can be grasped. But the extent of this dependency seems to me to be exaggerated. In discussion groups risks are taken in jumping from one concept to another related one without all the intermediate steps being filled in systematically. The gain is that the immediacy of people's interest is captured; the loss in logical order is often mitigated when the original sequence is picked up a bit later, perhaps with greater keenness to understand the intermediate steps.

This effect of inflexible sequences on the motivation of students is highlighted in an important report to the Australian government on re-imagining science education:[v]

> What is clear from the literature … is that the problem with student attitudes towards … school science relates to the transmissive and limited pedagogies used … science [needs to be positioned] for all students, as a way of understanding the world

and engaging with issues that are meaningful to them, and needs to move beyond restrictive notions of sequential conceptual understandings.

It's for geeks

Given the challenges set out above — obscure language, endless abstractions and too many specialist subjects — who can be blamed for thinking science is just for geeks? The portrayal of scientists in fiction as single-minded, asocial individuals with a mission doesn't help. Does this stereotyping spread beyond the professionals to the ordinary people with an amateur interest in science? Do they come across in a similar way — as fanatical about rare species of butterfly or the gigabyte capacity of a laptop? Fortunately, the evidence points in quite a different direction. As mentioned previously, not only do two thirds of people in an Ipsos MORI survey think "it is important to know about science in my daily life" but half (51%) say they "hear and see too little or far too little science" — an increase of 17 percentage points since 2008.[vi] The public at large, not just the geeks, seem to be increasingly interested in science.

Members of the discussion groups have extremely varied interests — journalism, marketing, psychotherapy and the performing arts, for example — and by no stretch of the imagination could they be described as "geeks". Discussion topics range from animal psychology through cloud formation to the nature of dark matter. As we saw in Chapter 2, fascination may start from any aspect of life, and is not restricted to the numerical or technical. One of the consequences of encouraging adults to talk about science freely, from experience, is discovering how easily issues of science emerge not only from technical but also social, political and religious points of departure, and how comfortably such scientific considerations lead back into public affairs. The breadth and variety of such discussion threads can be seen in the quotations and examples given later in this chapter.

It's to do with technology and engineering

For many people the distinction between engineering and science appears to be neither clear nor a matter of great interest: both are technical, practical and mathematical — and hence rather remote. A survey conducted for the Royal Academy of Engineering[vii] demonstrates just how low awareness of the nature of engineering and role of engineers is, with 59% of respondents agreeing that "hardly anyone knows what engineers do". Given this, it is hardly surprising that the distinction between science and engineering is not widely understood. The same survey reports that 51% of respondents agreed that "engineers are very similar to scientists".

In discussion groups I am often badgered light-heartedly about engineering issues: "Come on, Andrew, you're a scientist: how do I stop the pilot light blowing out on my boiler?" Although said half in jest these kinds of request reveal the extent to which the realms of science and engineering are merged in people's minds. As a science teacher with a mission for wider public understanding, I would be happy to respond to such a question with talk about the nature of gases and the concepts of combustion and chemical reaction. But this would hardly meet the questioner's immediate need for an engineer with a practical knowledge of boilers! Wikipedia describes the two fields succinctly:

> Engineering is the application of scientific, economic, social, and practical knowledge in order to invent, design, build, maintain, and improve structures, machines, devices, systems, materials and processes.[viii]

> Science (from Latin *scientia*, meaning "knowledge") is a systematic enterprise that builds and organizes knowledge in the form of testable explanations and predictions about the universe.[ix]

The two are of course intimately linked. Scientific discoveries are routinely exploited by engineers — advances in materials science for example enable lighter planes to be built and more fashionable clothing to be designed. Engineers, on the other hand, constantly present scientists with new challenges for research and theorisation, based on their experience of practical applications.

Distinctive as the two fields are, the questions people ask about the world lead into both, and move back and forth between them. A discussion about mobile phones for example led into both the science of electromagnetic waves, with their extraordinary ubiquity, frequency and invisibility and the technology of the rooftop masts and hand held receivers that make up the global network (see example on the next page). Interestingly, this same discussion also led into health risks, trust in scientists, communication habits and globalisation. So yes, science links readily to engineering, but in a free-ranging discussion it can also lead into social, economic, political and religious matters too.

The Interplay of Science and Engineering

A discussion about mobile (or cell) phones

"How does my particular mobile phone know to ring when someone calls me? How does it get picked out from all the others? Come to think of it, what is it that does the ringing anyway?" Thus began a lively discussion about the science and technology of mobile (or cell) phones. The thousands of telephone calls, all competing for attention in the room, the street, the neighbourhood, give you a sense of just how extraordinary the space around us is, filled with invisible signals aimed at all the phones, radios and TVs.

The signal for your phone radiates out from a mast somewhere near you, probably on top of a building, along with thousands of others at the same time. A special electronic code, unique to your particular phone, prefaces each signal and your phone is "listening out" all the time for a signal headed up with your unique code and ignoring all the others. When it does eventually detect your code, the phone starts ringing and prepares to receive the rest of the signal.

"That's all very well," someone interjected in the course of this explanation, "but how on earth does it know where you are — you move around don't you?" The country is divided into a set of small cells, each only a few miles across. With a mast in each cell, the unique signal that your phone is continuously bleeping out gets picked up by the nearest mast. Having located the cell you are in, anyone calling you is quickly connected to your nearest mast which sends its ring signal out to you.

"Are you saying the system knows where I am all the time?" was the next, faintly paranoid question. Yes, we all sacrifice a bit of privacy when our phone is switched on. If you are thinking of committing a crime, best to switch off your mobile before you start!

Armed with this basic information about the cellular system a more fundamental question followed. "Let's roll back a moment — are we saying there are signals buzzing all around us all the time? Are we living in an ocean of messages flying past in all directions?" The underlying nature of telephone signals needed to be explained, the electromagnetic fluctuations, vibrating a billion times every second emanating from aerials dotted around the country.

"But how do these electromagnetic waves give rise to the sound of a distant voice in your ear?" "What is sound anyway? How does the ear respond to it?" An avalanche of questions to feed many months of discussion.

Common Perceptions

Having considered something of how science is commonly perceived, we now look at some of the ways in which associations, both positive and negative, are made between science and everyday concerns. Some of the broad links that have emerged in discussion groups over many years of discussion are explored below.

Science for health care

Health is one of the aspects of life that inspires most questions in discussion groups. In general, science is perceived as vitally important in helping to understand medical conditions and improve treatments. Interest in understanding, scientifically, how nerves work, what hormones are, how radiation affects the body, and so on, is very high. Implicitly, science is seen in a positive light in relation to health.

However, the limitations of Western medical approaches are also a source of interest. Many people are not persuaded that the scientific procedures for creating, testing and administering drug treatments are the whole story. Interest in homeopathy, herbal and Chinese and Indian approaches is also high. Scientific dismissal of treatments for which randomised controlled trials show no evidence of effectiveness is criticised by some. The reductionist basis of the natural sciences is seen by some as missing out vital holistic aspects of good healthcare. This is a regular source of debate in groups, with firm views on both sides, and often leads on to discussion of the essential nature of the natural sciences. When it comes to the wider application of science in healthcare, there is less controversy and considerable interest — cancer treatments, radiography, endocrinology and mental illness, for example, have inspired dialogue about cell signalling, hormones and nerve transmission, as well as visits to labs and clinics.

Technology in the home

Science is appreciated for its role in providing the equipment that makes life so much easier for people today. The fridge, the telephone, central heating and computers are all sources of questions and discussion. The scientific ideas that underlie them prove fascinating and the benefits of learning about them are plain to see. Fundamental concepts, such as latent heat, electromagnetic waves and binary arithmetic develop naturally from considering the scientific bases of these devices. But even here critical attitudes to science sometimes emerge. Fears about the impact of electromagnetic radiation on the brain from mobile phones and about the intrusion of computer technology on children's play often flow from discussion of these scientific advances. These can provide starting points for discussion about the nature and quality of scientific evidence in relation to contemporary concerns. In some cases, such as the effects of electromagnetic radiation on the body, research evidence has not had time to build up, although public concern and personal anecdotes are widespread (see example in Chapter 9, page 177).

Science is amazing

Sheer amazement is often the reaction to developments in surgery or computing or space exploration. TV documentaries, news stories or popular science books inspire people, particularly when they emphasize extremes — minute keyhole surgery, massive computing power or remote exploration on Mars, for example. Technological triumphs of this kind make a big impression and encourage positive attitudes towards science. But in themselves they may not bring people any closer to the scientific ideas that underlie the technological achievement. However, discussions that start from everyday issues can lead on to scientific concepts that are equally amazing — in a different way. The idea that the air we breathe is made up of molecules buzzing around at hundreds of kilometres per hour or that the information needed to create our entire being is reproduced in the DNA in every cell of the body, has the potential to amaze, if broached carefully. Gaining insight into some of the fundamental concepts in science can be as exhilarating as the latest technological marvel.

But it's brought us terrible things

Almost no discussion about science concludes without reference to its darker side. The role of scientists and the unbridled pursuit of new knowledge remind people of the military consequences of atomic and nuclear research. Little could Madame Curie have guessed where her disinterested investigations of the radioactive mineral pitchblende would ultimately lead to. Of course people have been aware of the double-sided possibilities of scientific discovery long before the nuclear age — chlorine gas was used as a weapon of mass destruction in the First World War; science has always played a central role in military development. There was an era, up to the 1960s perhaps, in which public perceptions of science were less sceptical — a time when the British prime minister Harold Wilson could rouse public support in 1963 by arguing that a "new Britain" would need to be forged in the "white heat of this scientific revolution" and to write that his aspiration had been to "replace the cloth cap [with] the white laboratory coat as the symbol of British labour".[x] But attitudes have changed over the subsequent fifty years; such an appeal is less likely to be heard today. Fears over environmental damage, genetic modification and the safety of nuclear power plants, for example, have risen; some triggered by events, others by the precautionary principle: to veer on the safe side in the absence of clear evidence.

Scientists are unlike normal people

Stereotypical images of scientists are plain for all to see in film and on TV: asocial, obsessive and cold-hearted. In discussion groups it is clear that people know these to be caricatures; but it is less clear how scientists are actually perceived by the broad run of citizens. In an interesting study children were simply asked to draw a picture of a scientist. Lab coats, eyeglasses, facial hair and laboratory equipment were regularly occurring features. Only girls ever drew the scientist as female, in a sample of several thousand children. The study showed that this stereotypical imagery is already apparent in six- and seven-year-olds and strengthens with age.

People in discussion groups generally have little direct contact with scientists in their everyday lives. Does this lack of personal acquaintance contribute to the perception problem? Most of us have contact with shopkeepers, builders, teachers and nurses, but few of us meet with technical people other than in hospital or the computer store. Yet scientists are involved in very many routine activities that affect our daily lives, from water treatment through wildlife management to software development. In discussion, the images of scientists that occur most frequently are of white-coated research scientists in laboratories. When scientists are mentioned as individuals in discussion, the reference is usually to an exceptional contemporary figure, often a celebrity such as Stephen Hawking, or to a scientist who made an exceptionally important contribution in the past, such as Galileo or Darwin. The ordinary concrete technologist who travels to work in suit and tie rarely makes the news!

Visits to scientists in their working environments are a very helpful way of getting to understand the routine aspects of most scientific activity and the ordinary human characteristics of most scientists. Fortunately, there is now pressure on scientists to open themselves up to the public, to contribute to "public engagement", so it is becoming easier to organise such visits.

A Different Way of Seeing Science

We have now considered from several points of view the central paradox about engaging with science. In Chapter 2 we outlined some of the ways in which adults find themselves drawn to scientific ideas by curiosity emerging from their everyday lives. In Chapter 3 we looked back at the experience of science in the teenage years, largely at school, and saw how it had proved unsatisfactory for a large proportion of people. We considered, in particular, how it had differed from other subjects at school and often led on in adulthood to a disproportionately weak grasp of scientific ideas. We now draw on the experience of discussion groups that use a quite different approach to suggest ways in which science can be perceived differently.

Scope

The Ipsos MORI survey of attitudes to science mentioned earlier in this chapter[xi] found that most young people think of the sciences in terms of three subjects — biology, chemistry and physics — although older people tend to emphasise the aspects of science that are most prominent in their own lives such as health, drugs and cures for diseases. The experience of discussion groups, over many years, confirms the latter point and reveals just how wide the range of topics can be when discussion starts from everyday concerns and is allowed to range freely. Within the natural sciences, discussion ranges over disciplines such as cosmology, nanotechnology, digital communications, endocrinology, mental health and experimental psychology as well as the traditional branches of physics, chemistry and biology. The boundaries between subjects are crossed time and again in discussion as a question is pursued through all its ramifications. Here are typical examples of paths followed in discussion groups:

Example 1

What are waves? → graphical representation of a wave → graph of oscillations → recollection of oscilloscope screens in hospital → wave nature of sound → function of the ear → nature of electromagnetic waves → function of a microphone → digital reproduction → maths of binary digits → digital transmission

Example 2

News about the Worldwide Telescope → scientific meaning of resolution→ dots on a TV screen → digital representation → CD re-mastering → ultra-violet radiation in a CD player → mixture of colours → the ear and sound → what musicians do → expert performance → nature of savants → effects of damaged brains

What starts with a reasonably defined scientific question (about waves or telescopes in these two cases) leads on to points of reference in people's own experience: oscilloscope traces recollected from hospital, coloured dots on a TV screen, a re-mastered CD. The path of enquiry may then deepen as expected in a conventional linear syllabus or it may jump by an analogy that seems appropriate to the participants, though it might well be off-limits in conventional subject-based learning. Thus, in the first example the path from oscilloscopes to sound waves and on to the function of the ear is a perfectly familiar one, following a logical path; though in a formal educational programme even this would present practical problems as waves might be taught in a physics lesson and ear structure in biology — the intimate connection between the two might be missed. In the second example, a relatively conventional move from telescopes to resolution (the capacity to distinguish fine detail) then shifts to dots on a screen and digital representation in general. Then, by analogy, it jumps to CDs, ears, musicians and eventually to the brain. In this kind of discussion, no individual topic is delved into deeply, but the interest lies in the numerous connections made to aspects of everyday life and to the underlying unity of digital representation, visual and aural.

The range of subject areas drawn into a discussion can easily extend beyond the main scientific disciplines. Related subject areas, in which natural sciences play a partial role, such as archaeology, ceramics, anthropology or criminology, feature regularly in the course of discussion. Experiences from everyday life often lie within one of these realms and may act as starting points for enquiries that lead on into fundamental scientific ideas. A discussion initiated by a museum employee about dating ancient objects, for example, led into consideration of carbon-14, which in turn stimulated questions about radioactive isotopes and ultimately led into the fundamentals of atomic structure. Similarly, discussion about glazes, initiated by a potter, led into the physics of temperature and the chemistry of reactions. The key point about these examples is that the direction of flow is reversed. In place of a sequence of steps in physical or chemical theory leading to a prescribed conclusion, from which a link to everyday life is made, the real-life situation with its burning question leads on to an unpredictable encounter with deep theory.

Discussion may start from an issue apparently outside the scope of the natural sciences. Two examples of actual discussion paths show how a current social issue can lead into science. The first began with a conversation about a gay male friend and ended with the science of statistics:

A gay male friend → are gay men antagonistic to women? → influences on behaviour → cultural studies vs biology → individuals and populations → population statistics → distribution graphs → verifying trends

The second example began with the topic of psychotherapy, specifically: why does a course of treatment take so long?

The book *Why Love Matters* → biological vs environmental factors → Freud's concepts → cross-cultural differences

At this point, the discussion became so lively that a lady sitting at an adjacent table in the wine bar simply couldn't resist joining in. She turned out to be an Alaskan anthropologist full of relevant information about the sex lives of Aleuts. So the discussion took an unexpected turn:

Aleut wives sleeping with visitors → increase gene pool → marriage taboos in different cultures → cultural norms → learning vs instinct → long infancy in *Homo sapiens* → longitudinal studies of early experience on later life

So when we start from real issues in people's lives, discussion crosses freely and regularly over the boundaries within which subject disciplines are organised. This may well present difficulties for the orderly presentation of well-established scientific knowledge, but it can also result in strong motivation to persist in pursuing scientific ideas.

It is also common for discussion to lead to issues that lie outside the natural sciences (in philosophy, religion or the social sciences, for example). One session began by remarking on the enthusiasm shown for science by primary school children, then led on to the division between vocational and academic education. This led in turn to Descartes' separation of mind and body, and differences between analytic and continental approaches in philosophy. In another case, discussion triggered by Damasio's book on consciousness centred on experiments with three- and four-year-olds about their ability to empathise, to understand another person's selfhood. Freud's methodology was discussed and led on to the pros and cons of positivist philosophy. Discussions of this kind permit participants to contribute from their wider experience and reading, while feeling they are also engaging with scientific ideas.

Discussions may well start from issues that lie in both the natural and the social sciences, such as autism, mental health or care of the elderly. Insights from the natural sciences — into the nervous system, brain function and human development, for example — sit alongside social science insights into behaviour modification, therapies and care regimes. As the discussion group is science-based this means that scientific aspects of social issues that may otherwise be overlooked are emphasised. Experimental psychology studies of infants or longitudinal studies of people's lives, for example, not only inform us about social and economic matters but can also introduce concepts of statistical and experimental methods, demonstrating the continuity of science across the natural and social domains.

Uncertainty

The twentieth-century painter Georges Braque famously said: "Art disturbs, science reassures." Whether this is always true of art I am not sure, but this description of science is harder to sustain today. Uncertainty about the local effects of global warming or statistics about the likelihood of inheriting a genetic disorder are part of science today and aren't entirely reassuring. Uncertainty lies not only in the quality of information that science provides but also in the way scientists go about their work. The much respected zoologist, J.Z. Young, said in his 1950 Reith lectures:[xii]

> I have often thought that one of the characteristics of scientists and their work is a certain confusion, almost a muddle. This may seem strange if you have come to think of science ... as being all clearness and light ... his [*sic*] work is a feeling out into the dark, as it were. When pressed to say what he is doing he may present a picture of uncertainty or doubt, even of actual confusion.

Developments in quantum physics add a further twist to the reassuring idea of certainty in science. Progress in theories about the ultimate nature of matter is leading to less and less certainty about the meaning of fundamental concepts. When matter is dissected further and further does it remain matter, or simply the vibrations of a notional string? When time is rolled back to the earliest days of the universe does time itself have meaning? Were time and space themselves created with the Big Bang? If it's reassurance you are seeking, I am not sure I would advise science as your first stop.

In a conversation in the "online salon" *Edge*, the theoretical physicist Carol Ravelli says:

> What I see as the deepest misunderstanding about science ... is the idea that science is about certainty. Science is not about certainty, it is about finding the most reliable way of thinking, at the present level of knowledge. ... Not only is it not certain, but it's the lack of certainty that grounds it. Scientific ideas are credible ... because they are the ones that have survived all the possible past critiques.

So science is shot through with uncertainty — in its finding, its theories, its methods and in the minds of its actors. How does this compare with popular perceptions of science? How does science, seen in this light, compare with other competing subjects, especially at school? What does it tell us about the possibilities of "science for all" emerging as a branch of the humanities, accessible to all? The issue of uncertainty is explored further in Chapter 9.

Learning about science itself

In everyday life we cope with the fact that we simply don't know, with any degree of certainty, much about things. The causes of the First World War are disputed and, for Mao Tse-tung, even the French Revolution is too recent for its impact to be assessed. To decide which school our child might be happiest in, or what to do to relieve our back pain, we gather what information is to hand, listen to the experience of our friends and rely on some kind of "gut feeling". Indeed, in many situations, not only do we lack full understanding, we are also unsure how to judge whether our decisions were wise. In retrospect, in choosing a school for our child were we right in prioritising happiness or academic attainment or future financial security?

On the whole, the world that science deals with is much the same. We have known for many decades that an aspirin will relieve headaches in most people most of the time; but understanding the precise way in which it achieves this came relatively recently. The substance that was later named "aspirin" was first made in 1897 and shown to be effective as an anti-inflammatory analgesic drug. It took a further seventy-four years to discover the mechanism by which it exerts its effects: inhibition of an enzyme that makes a substance that causes inflammation, swelling, pain and fever.[xiii]

In fact, the realm of the "known unknown" is not only huge but is expanding as scientists probe more deeply into things. The history of cancer research, for example, shows how understanding can grow and treatments improve, but at the same time how new, unpredicted questions arise, calling for further research, as more and more is discovered about the workings of the cell and its environment. However, the impression of science that we pick up at school and, to some extent the media, is of a special world of cast-iron certainties. Indeed, the historically favoured use of the word "law" to describe the mathematically exact relationships in science encapsulates this sense of certainty and tends to promulgate it.

So if mathematical laws and absolute certainty fail to capture accurately the nature of science as it really is, what concepts belong in their place? The sense we get of scientific progress, as a series of amazing breakthroughs made by outstanding individuals, is just part of the story. The philosopher of science, Thomas Kuhn, showed that most scientific activity is "normal", with scientists following the prevailing theory, collecting data and looking for patterns in it. We build up insights and understandings from these patterns. This could be said of many branches of the humanities and social sciences too — history, geography and sociology, for example — in which data are often taken from direct observation, as in areas of science such as astronomy or geology. What is distinctive about many (though by no means all) of the natural sciences is the opportunity to set up experiments. In these situations, the complexity of some real-world situation is made manageable by constructing an arrangement in which only a limited number of variables are believed to be in play. This simplification enables separate factors, each of which might be important in a situation, to be disentangled.

Take, for example, protein molecules that have flexible structures comprising hundreds of atoms that flail about in a minestrone soup of other molecules. Flexibility and movement are essential for the molecules' function, but make for difficulties in studying the molecule structures precisely. This is best done by isolating millions of identical proteins and preparing them in a single crystalline form. In this constrained condition the precise structure can be determined using X-rays, and from this its function and interactions may be inferred. Thus, important characteristics of the protein are constrained in the experimental set-up, in order that other features can be studied effectively. Constraining reality in experiments such as these is justified because the knowledge gained is useful, rather than because it is complete. Understanding protein structures, for example, can lead to new medicines to combat disease.

The purpose of this example is to illustrate a fundamental idea about the nature of science that is easily overlooked: science creates *models* of reality, fitted to some purpose, rather than describing reality itself. This pragmatic (and humble) view of science is relatively modern and contrasts with the more straightforward idea, strengthened in the eighteenth and nineteenth centuries, that observations of the physical world result in a complete and final description of an objective reality.

Historically, many theories that were once believed to describe the reality of the physical universe exactly were later shown to have limitations and were superseded. Newton's highly elegant theory of gravitation, precise enough for route planning to the Moon, gave way to Einstein's theory in the twentieth century, with its greater precision over astronomical distances. The concept of light as a form of wave, developed in the eighteenth century, had to yield to a more ambiguous concept when its quantum nature was also revealed at the beginning of the twentieth century. Experiments, designed deliberately to probe specific aspects of the physical universe, provide data upon which scientists are able to base their models. An experiment in 1887, showing that a mythical "aether" permeating all space didn't exist, was a foundation for relativity theory; similarly, experiments with electric currents created when ultraviolet light shines on a metal surface heralded the quantum theory of light. Given that what scientists spend their time doing is trying to interpret data and create models, it is hardly surprising that these interpretations and models are often hotly contested. In the early development of new ideas there's plenty of room for discussion and interpretation. Looking back on the historical development of science one is forced to understand knowledge as provisional — much of it is destined, ultimately, to become outdated or at least refined.

In addressing issues such as gravitation or quantum effects with adults, the thread of discussion regularly moves into questions about the nature of reality and the role of science. It is clear that by introducing the ideas of interpretation, contestation and provisionality in science, people's understanding of the nature of the subject can be transformed. By bringing out the more uncertain aspects of the subject, to complement its more definitive mathematical aspect, people see it as more closely related to other subjects in the arts and humanities. As a result, curiosity develops from the original questioning about phenomena in the world to deeper issues about the nature of theory and experimentation.

In Chapter 9 we examine ways in which the nature of science itself can become the topic of discussion. Reflecting on the nature of a subject as well as its content is of course quite familiar in other fields, such as literature or history or sociology, where we become aware that literary criticism or historical analysis also involves exploration of what it is to criticise or analyse. Encouraging exploration of the nature of science — experimentation, model building, falsification — has the potential to stimulate curiosity about science and motivate people to engage with it. It not only offers a richer understanding of the subject, it also makes it more attractive to a wider audience.

Understanding scientists

We know something about public perceptions of scientists from the Ipsos MORI survey[xiv] mentioned previously. The 2011 survey showed that scientists are generally perceived as serious, objective and rational, and are unlikely to be perceived as friendly or "good at public relations". In discussion groups great scientists of the past are occasionally discussed especially if their personality or behaviour seems remarkable. The extremely meticulous observations by Tycho Brahe in astronomy and Gregor Mendel in genetics over many years, or the extraordinary imagination of Mendeleev in visualising the periodic table, fascinate through their human appeal. But the character of ordinary scientists milling around in society today rarely enters discussion. However, perceptions of scientists as being more like you and me are beginning to develop thanks to the popularity of TV programmes fronted by media-friendly practitioners. Visits made by discussion groups to scientists in their laboratories tend to reinforce this more realistic perception of scientists as much like the rest of us — witty, long-winded, fascinating, altruistic, self-centred — a mix of human characteristics.

However, it is worth considering for a moment some special characteristics of their background and training that do distinguish them from others. In England they will mostly have specialised in science subjects since their A-levels and mostly had to cease studying the arts after the age of sixteen: no literature, art, history, sociology or languages. In addition, many of them will have had to take mathematics alongside their science subjects, restricting their options even further. After studying at degree level, again mostly with choices restricted to science topics, some will have moved into further specialisation for several more years at doctoral or professional levels. Their training is long and increasingly specialised. The chances are that most scientists will have had little opportunity to develop their interests very broadly across the humanities and arts, however willing they may have been. In short, those who ultimately become practising scientists (in England at least) are in effect survivors of a particularly narrowing education process.

To have survived such a lengthy training process suggests that, whatever it is that motivates a person in science, must be strongly felt. Motives are very varied and overlap considerably with creative people in all walks of life — desire for explanation, pleasure in reasoning, tendency to seek patterns, wish to advance knowledge and improve the world. Interestingly, a study on the personality of science researchers[xv] shows that:

> Compared with [the] general population researchers appeared as generally intro-verted (withdrawn, internally preoccupied, inhibited and self-sufficient) … creative people in art and literature shared a similar profile to that of researchers, so intro-version seems to be characteristic of creativity in general, rather than science *per se*.

Whatever part altruism plays in the personal motivation of scientists they are perceived positively by the public in this respect. In the 2011 Ipsos MORI survey, "nine in ten (88%) think *scientists make a valuable contribution to society* and eight in ten (82%) agree they *want to make life better for the average person*" (emphasis in the original).[xvi]

Scientifically trained people work in a huge variety of occupations from food technology to music recording, and animal breeding to blood testing. Many perform regular routine procedures, such as checking for bacteria in food or measuring the strength of materials. Many are involved in designing and testing innovative products, others in research or in science-based practices such as medicine or engineering. Some go on to teach the next generation. Like other researchers, scientists work in a highly competitive environment, fighting for limited funding and exposing their work to "peer review" — the procedure for evaluating each others' proposals and articles. In recent decades scientists have been encouraged to extend their repertoire by engaging more closely with the public to explain their work and combat the rise in negative public attitudes to science. A small "public engagement" industry is developing to facilitate this through festivals, open days, consultation exercises and the like. From the point of view of ordinary members of the public this is a huge step forward and represents an unprecedented opportunity to start getting to grips with science. Further thoughts about these opportunities are offered in Chapter 8.

A Different Way of Learning Science

In this chapter we have looked at how science can appear to people who have had little contact with it since school and we have considered how it might be approached in a quite different way. Insight into these issues has come partly from studies of attitudes to science and partly from the experience of discussion groups that work in an alternative way. We now conclude the chapter by considering how these discussion groups demonstrate the possibility of an alternative way of engaging with scientific ideas.

An international study of science education provides a useful starting point. The ROSE project (Relevance of Science Education)[xvii] surveyed young people in dozens of countries about their attitudes to science. It concludes:

> Teaching has … to be motivating, meaningful and engaging. It has, in some way, to link up to the values and interest that the learner brings to the classroom. If not, no other "learning" than rote memory based on duty is likely to occur. If this is the situation, the learner is likely to develop negative attitudes, and will turn their backs to SMT [science, maths and technology] when they make their decisions about future life, be it as students or as citizens. … Students' own attitudes, values and interests should be given high priority in the selection and presentation of the science curriculum contents … Teaching practices that do not engage students in meaningful learning are not likely to give lasting positive results.

These conclusions reflect much of what contemporary theory has to say about effective learning, not only in science but in any subject area, and emphasise the importance of starting from where the learner is at. A review of research on science learning in the primary school by the National Foundation for Educational Research[xviii] reinforces this point:

> The main thrust … in science is the need to find out what pupils know, what they don't know and what they partly know — their misconceptions — and to develop teaching that will move their understanding on. The [research] literature explores the need for a range of questioning, the importance of talk and discussion and the provision of appropriate feedback.

The discussion groups set up by the author in adult learning centres and informal surroundings were designed with such approaches in mind. In these groups, issues for discussion are simply put forward by members of the group, on the day. As we saw in Chapter 2 they may arise from current affairs, the weather, things happening at home or almost any observation about life. The key point is that starting points arise from their own questions. Then, in place of an "expert" attempting to answer the impromptu questions, everyone is encouraged to express an initial view or to bring up any experience or recall knowledge that seems relevant to them. This leads to an exchange of ideas between participants, sometimes challenging each other's thoughts, sometimes reinforcing them by offering comparable experiences. Lines of thought are pursued and new questions are raised as the group strives to think things through for itself, reasoning its way through plausible explanations, inconsistencies and dead ends. For the members of the group this approach contrasts starkly with their initial expectation of getting a direct answer to a direct question. Equally it offers a tough challenge for the tutor: to hold back from answering, to stand by, encouraging the group to continue its exchange of thoughts as far as it will go.

As the limits of peer discussion are reached a strong appetite develops for the tutor to intervene. Sometimes the group's demand for explanation seems as intense as it is for children. But in practice, of course, the tutor may or may not be able to help with the precise items of knowledge requested. "What is quicksand?" "Is lactose intolerance ethnic?" "How do children learn to pitch notes?" This selection of questions from one two-month period alone demonstrates the impossibility of a single tutor having all the fingertip knowledge required. But simply providing answers to questions is not the purpose of the exercise; there are plenty of websites available for that. The aim is to help people engage with fundamental ideas in science. The question, however pressing it may seem, is simply a way in. So the tutor's task at this point is not to answer the question directly, but to identify the underlying scientific concepts in play and to relate the query to the appropriate concept. Thus, quicksand might lead into the composition of materials or density and flotation; lactose intolerance into chemistry of sugars or allergy and the immune system. Again the direction of travel depends on how the group picks up on the newly introduced concepts.

At this point people sometimes return to the original question to pursue it in greater depth, but often they stay with the newly introduced concept as they wrestle with the intellectual challenge it presents. In a memorable example, one discussion group began with a desire to know more about Einstein's theory of special relativity. After batting around various points people had picked up about the theory, the tutor introduced the classical, *pre*-Einstein theory of relativity as a preliminary concept. By introducing a real-life issue — when to throw a used drink can into a pavement bin as you pass by on a bike — it became clear that people disagreed about the basics of relative motion. So this, the classical seventeenth-century concept of relativity, became the point of argument and resolution — a productive and manageable moment of learning and also a necessary prelude to any subsequent discussion of Einstein's theory.

So, as it transpires, although many questions from ordinary life lead into relatively complicated areas of science, the discussion that ensues can centre on more basic scientific concepts. One of the most interesting discoveries in running open-ended discussion groups is just how fascinated people can become with apparently mundane topics of science that might have previously been seen as dull. For example a discussion about salting the roads in icy weather led on to asking why icebergs float mostly below the surface. Floating twigs and logs were compared, leading on to arguments about the meaning of density, weight and volume and ultimately to the concept behind Archimedes' famous principle — a fascinating, animated discussion drawing repeatedly on everyday observations, revolving around relatively ordinary concepts (see *Expressing Your First Thoughts* in Chapter 7, page 131).

After the introduction of relevant scientific concepts and the challenges and debate they provoke, further questions arise, often more refined ones. The process of exploring prior understanding, exchanging experiences and debating possible explanations then begins again; the process is iterative. However, at some point, the options for learning from the group itself become exhausted and a thirst for expert knowledge becomes unstoppable. At this point choices arise. It may be that the tutor's background fits the topic, so she or he is able to give a small talk, pulling together strands of discussion and explaining in a relatively conventional way the scientific ideas. More often the tutor's role is to formulate in a scientifically informed way the knowledge that is being sought and to suggest ways of finding it. An example of the various stages in the process is given below.

The internet provides a simple, accessible way for anyone in the discussion groups to pose questions and pursue an enquiry. Popular science books are plentiful and useful but are less easy to make use of in relation to a specific question. The experience of discussion groups over many years reveals a number of difficulties people encounter in using either of these kinds of resource; these are explored in Chapter 8. A key role for a scientifically trained tutor is to look for resources on the internet and select ones that appear relevant, sound and readable. Good material on the internet is buried amongst a much greater quantity of poor material so it is difficult for

The Stages of Discussion

Exploring the meaning of energy

The apparently simple question "What is meant by energy?" was once asked in a discussion. After an unusual period of silence the group began to offer their preliminary thoughts and ideas soon flowed freely. "It can be felt but is not tangible — like the heat of the Sun," said the first contributor, and "It's a force of some kind," said another — both keeping fairly close to the scientific notion. Other responses were more socially based: "It's when you feel something about somebody — an intuition"; "People give out positive or negative energy vibes." The link between energy and the body was touched on: "Food gives you energy. You never have enough energy, you need a lot of it." And more interestingly: "Maybe thought could be energy or give rise to energy. Are brainwaves energy?" With the widespread use of the word in alternative health practices the question was asked whether only a Western definition was permitted. "When I learned at school that energy is neither created nor destroyed I thought this is consistent with the Eastern view of energy being transformed," said a person with roots in India.

After this exchange of thoughts about the word, the moment arrived for some input on its scientific meaning — a word used across a vast range of disciplines, from engineering and astronomy to physiology and sports science, yet defined in a universal way. Its definition as "the capacity to do work" led on to explanation of the meaning of the word "work" in science and the related concepts of force and distance. Other connotations of energy were introduced: energy in its many forms — in heat, light and microwaves, in electrical flow and mechanical movement, in chemical bonds and atomic nuclei — and energy as an abstract entity constantly being transformed in car engines and living bodies.

These explanations gave rise in turn to further questions: "Where does energy go?" "How does the body get energy out of food?" "Is it the same as power?" More references were made to personal experiences: "What is happening when milk continues to boil after the gas is switched off — does it have a kind of energy?"; "There are different speeds of heating up — in pottery Raku firing is slow compared to normal."

Further scientific input followed, touching on common examples of energy transformation, like heat dispersing from your body or a heated

(Continued)

> **(*Continued*)**
>
> room or the Earth itself. Discussion moved on to deeper concepts: the inexorable running down of the universe and the concept of entropy as a measure of disorder. The session ended with thoughts about the use of words like energy in both common parlance and scientific contexts. The scope of science with its focus on objective aspects of reality, like the energy content of food, was contrasted with the wider world in which the notion of energy is often associated with subjective feelings.

untrained people to find it. In this way responses to specialist questions can be made in a subsequent session, *after* the broad issue has been teased out in the first discussion. Thus one discussion may build on the previous one as more specialist information is gathered.

But for a truly expert view and real insight into how science actually works, nothing is more inspiring than a visit to a scientist in their laboratory. Visits not only mean your questions can be addressed authoritatively, but also the opportunity of dialogue means that the specific obstacles to understanding can be tackled on the spot. Experience with discussion groups suggests that scientists tend to be only too willing to stay behind after hours to describe their work, show how their research is carried out and to tackle some of the tricky questions ordinary punters are inclined to ask. Cancer specialists, earth scientists, particle physicists, endocrinologists, MRI operators and musical performance scientists have all proved ready and willing to engage with groups, and have often found the experience rewarding.

Summary

In this chapter we have outlined what science learning might look like, were it to be detached from the training of future scientists. We have considered how it appears to the majority of people who left science behind at sixteen — perceptions of its difficulty, its remoteness, its reliance on mathematics, and so on. We have looked at the kind of attitudes that grow up around science, especially for those who have little direct contact with it. More positively, ways in which science learning might be reframed for the majority have been sketched and examples given of how this works in discussion groups organised for the non-specialist. In the next chapter we draw on the experience of hundreds of actual science discussions to look in greater detail at what it is that excites our curiosity, how it may be dampened and how it may be sustained.

Endnotes

i.   Millar, R. (1991). 'Why is Science Hard to Learn?' *Journal of Computer Assisted Learning*, 7(2), 66–74.

ii. Royal Society (2008). *Science and Mathematics Education 14–19.* Available at: http://royalsociety.org/education/policy/state-of-nation/14-19/.

iii. Coe, R., Searle, J., Barmby, P., Jones, K. and Higgins, S. (2008). *Relative Difficulty of Examinations in Different Subjects.* Durham: CEM Centre. Available at: http://www.cemcentre.org/documents/news/SCORE2008report.pdf.

iv. Barnett, M. (2000). 'The emergence of physics as an academic subject,' in Morris, A. (ed.), *Revitalising Physics Education*, Bristol: Institute of Physics Publishing.

v.   Tytler, R. (2007). *Re-Imagining Science Education: Engaging Students in Science for Australia's Future*, p. 12. Available at: http://research.acer.edu.au/aer/3.

vi. Ipsos MORI (2011). *BIS Public Attitudes to Science 2011*, p. 10. Available at: http://www.ipsos-mori.com/researchpublications/researcharchive/2764/Public-attitudes-to-science-2011.aspx.

vii. Royal Academy of Engineering (2007). *Public Attitudes to and Perceptions of Engineering and Engineers 2007*, p27. Available at: http://www.raeng.org.uk/publications.

viii. Wikipedia (2014). *Engineering.* Available at: http://en.wikipedia.org/wiki/Engineering.

ix. Wikipedia (2014). *Science.* Available at: http://en.wikipedia.org/wiki/Science.

x.   Francis, M. (2013). *Harold Wilson's 'White Heat of Technology' Speech 50 Years On.* Available at: http://www.theguardian.com/science/political-science/2013/sep/19/harold-wilson-white-heat-technology-speech.

xi. Ipsos MORI (2011). *BIS Public Attitudes to Science 2011*, p. 8. Available at: http://www.ipsos-mori.com/researchpublications/researcharchive/2764/Public-attitudes-to-science-2011.aspx.

xii. Young, J.Z. (1950). *Reith Lectures 1950: Doubt and Certainty in Science. Lecture 1: The Biologist's Approach to Man.* Available at: http://downloads.bbc.co.uk/rmhttp/radio4/transcripts/1950_reith1.pdf.

xiii. Vane, J.R. and Botting, R.M. (2003). 'The Mechanism of Action of Aspirin.' *Thrombosis Research*, 110(5–6), 255–258.

xiv. Ipsos MORI (2011). *BIS Public Attitudes to Science 2011*, p. 7. Available at: http://www.ipsos-mori.com/researchpublications/researcharchive/2764/Public-attitudes-to-science-2011.aspx.

xv. Wilson, G.D. and Jackson, C. (1994). 'The Personality of Physicists.' *Personality and Individual Differences*, 16(1), 187–189.

xvi. Ipsos MORI (2011). *BIS Public Attitudes to Science 2011*, p. 7. Available at: http://www.ipsos-mori.com/researchpublications/researcharchive/2764/Public-attitudes-to-science-2011.aspx.

xvii. Sjøberg, S. and Schreiner, C. (2010). *The ROSE Project: An Overview and Key Findings*, p. 29. Available at: http://roseproject.no/publications/english-pub.html.

xviii. Hodgson, C. and Pyle, K. (2010). *A Literature Review of Assessment for Learning in Science*. Slough: National Foundation for Educational Research.

Chapter 5

What Excites Our Curiosity?

Introduction

Having considered how science might be approached in a different way, building on curiosity and personal experience, we look in the chapters that follow at what happens in practice when groups of adults learn in this manner. We consider the kind of scientific issues that emerge — the curriculum as it were — and the manner in which they emerge. We start in this chapter by looking at the kinds of questions that arise when curiosity is given full rein. In Chapter 6, we consider some of the underlying themes that emerge from pursuing such questions over extended periods of time.

Curiosity

As parents know only too well, the curiosity of young children is boundless. Alison Gopnik, a leading authority on children's learning, states in her book *The Philosophical Baby* that "very young children are consumed by insatiable curiosity about causes, as their unstoppable 'why?' questions show".[i] She explains that earlier studies by Piaget, which suggested that children knew little about causation, have been superseded by more recent research. It appears the questions children were originally asked by Piaget tended to be hard and inappropriate, such as "Why does it get dark at night?" (Although this actually produced the child-logical response: "So we can sleep.") When asked in more recent research about things they were much more likely to actually know about — "Why did Johnny open the refrigerator when he was hungry?" — children as young as two gave sensible causal explanations.

This sense of curiosity about causation in younger children is reflected in their attitudes to science at primary school. A review of research on science education[ii] suggests that in most countries children enter secondary school with a highly favourable attitude towards science, as well as an interest in it. Unfortunately, this doesn't last — "much of an initial Yr. 7's (12 year old's) enthusiasm for science is dissipated over the five years of [secondary school]" in the words of an educational researcher who followed a group of children for the five years of their secondary education.[iii] We know that the switch away from science at school sets in during the early years of secondary education or, in developmental terms, early adolescence. A review of research[iv] on this question suggests that, for the majority of students, interest in pursuing further study of science has largely been formed by the time children are fourteen.

The switch away from science at school is not limited to the UK. A major international study, the Relevance of Science Education (ROSE),[v] states "for European countries and Japan, the answers [to survey questions] indicate that school science fails in many ways". According to the survey, most young people in these countries find science at school less interesting than other subjects and feel it has not increased their curiosity. Interestingly the same survey also reveals that, despite their negative views about science at school, young people have positive attitudes to science and technology in general. They tend to agree, for example, that "science and technology make our lives healthier, easier and more comfortable" and that "the application of science and new technologies will make people's lives more interesting".

For adults, surveys of the public's understanding of science and attitudes to it are conducted in both the US[vi] and in Europe. A US survey in 2004, by the National Science Foundation, revealed that many citizens "do not have a firm grasp of basic scientific facts and concepts, nor do they have an understanding of the scientific process"and that "the majority of the general public knows a little but not a lot about science". For example, most Americans know that the Earth travels around the Sun and that light travels faster than sound. However, few know the definition of a molecule. In Europe, the EU Eurobarometer survey in 2013[vii] showed that "the majority of Europeans (58%) do not feel informed about developments in science and technology". Yet despite this "at least half are interested in developments in science and technology (53%), with 13% very interested".

So, rigorous studies back up what is pretty well known anecdotally: most adults have little scientific knowledge. What the studies show that is slightly less well known is that both youngsters at school and adults at large, despite their lack of understanding, remain generally interested in science and positive about it. Anyone who has visited a bookshop in recent years can see plainly the growing interest in popular science books; today whole sections are dedicated to the category where once a few school textbooks or natural history titles might have been the limit. The rapid rise of science-based documentaries mirrors this, with whole series dedicated to both physical and biological topics — from volcanoes to space flight, evolution to surgery. A new cadre of celebrities has emerged to front this burgeoning field, often combining expert knowledge with skill in communicating effectively with the public.

This body of public evidence, both anecdotal and research-based, coincides with evidence from discussion groups run by the author over many years. The sheer fact that participants remain in the groups for many years (over a decade in some cases) shows how strong the desire for scientific understanding can be. It is natural curiosity that drives this interest and clearly curiosity, though easily quelled by over-formal instructional approaches, can equally be aroused under the right conditions.

In the following section we draw upon records of over 200 open-ended discussion sessions in which no particular scientific topic was billed in advance. On each occasion participants were simply free to open up on any topic they wished that seemed to have some relation to science. The starting point may have been a casual observation, a long-held question, a comment on a book or TV or radio programme or a practical matter affecting their lives. By reviewing these records a picture emerges of the kinds of questions people tend to ask and the many and various ways in which the thread of discussion evolves from them.

Sources of Curiosity

In Chapter 2 we surveyed the broad areas of life that tend to stimulate questions for discussion. We now explore these in greater depth using examples of actual discussions that have taken place.

The world about us

A major source of questions is the natural world around us. Observations of the nature of clouds, the behaviour of the sea or the causes of weather events are regular starting points for discussion. Is it the sense of mystery about these aspects of our surroundings that stimulates questions, or perhaps their ubiquity (that they are open for all to see and with us every day)? Let us look at some examples of what triggers off questions of this type and see how they so easily lead into deep science.

On one occasion a person had been on a summer visit to a seaside resort in north-west England and noticed that, bizarrely, the beach doubled as a car park. She asked, "Why is the sea so far out in summer at Blackpool?" Others in the group tried to puzzle out why the tide might be so low in summer. Could it be the extra evaporation due to the warm atmosphere? Perhaps it was the heavier rainfall that brought the level up in the winter? One person suddenly recalled that the polar ice cap melts in summer — might that explain it? Then another pointed out this would cause a rise in sea level rather than a fall. The analysis deepened as a third protested:

> But planet Earth is not like a glass of water with ice cubes in. The water holds on by gravity and spreads out all over the surface. All the water is joined up, so if the level rises doesn't it rise everywhere?

After this interplay of preliminary ideas, scientific knowledge was introduced, with a reminder of the different ways in which water moves in the sea: the rise and fall of tides, the steady flow of currents and the battering of waves. This simple clarification reminded people of things they vaguely remembered hearing about: wasn't there a huge tidal surge that ran up the Bristol Channel? Doesn't Southampton have double tides, due to the Isle of Wight? (Incidentally it does, but for a different reason.)

The concept of gravitational attraction was introduced — the Moon pulling up the water in the sea on the side of the Earth closest to itself, causing the sea level to rise beneath where the Moon is. Then the fact of the Earth's rotation was added in, causing the Earth to shift beneath this steady high point of sea level. So the reason why the time of high tide varies from place to place gradually became clear.

Further questions and challenges arose from these points until ultimately the fundamental Newtonian theory of gravitation was revealed: the powerful generalisation that any two objects attract one another, simply as a consequence of the matter they contain. By grasping this fundamental idea, other implications of the theory were intuited almost immediately: other planets must also have tides, they must be of different strengths and periodicities; the Earth itself, though solid, must be affected by tidal pull internally. Then even more penetrating questions followed — why do the tides recur every twelve hours if the Earth only rotates every twenty-four? Why are there four tides in Southampton? And so on. A chance observation in highly specific circumstances had led to insights into one of the most fundamental and general of scientific concepts.

In another discussion, a simple question was posed: why are sunsets beautiful? And in an unexpected rejoinder: is this connected to the red shift? This example illustrates how easily questions drawn from the world around us lead into aesthetic as well as scientific considerations. It also shows how a simple misunderstanding linking two unconnected phenomena — the red shift in astronomy and red sky in the Earth's atmosphere — can be brought to the surface in a relaxed, open setting where free questioning is actively encouraged. In this example, open discussion began with a mixture of loosely connected thoughts about colour:

> "Colour definitely affects your mood."
> "Do you remember how babies used to be dressed in pastel shades?"
> "Things are brighter these days … and they move faster than years ago"
> "Your aesthetic reactions are relative: you are content with your own house till you see a more beautiful one."

Soon this initial exchange of thoughts about aesthetics gave way to a demand for an explanation of why sunsets are red. This provided the opportunity to introduce two fundamental scientific concepts with immense power in explaining everyday phenomena: refraction (the bending of light at a change of medium) and dispersion (the spreading out of light of different colours).

Another discussion was triggered off by a holiday visit to Iceland where two tectonic plates are pulling apart. This led into enquiries about the theory of continental drift and the nature of tectonic plates. This in turn led to insights into the nature of the Earth's magnetic field and how the traces it leaves in the bed of the sea show that the direction of the field has reversed over geological time. Discussion took a more political turn when one contributor mentioned the possibility of burying radioactive waste in the "subduction zone" — the area where one plate is slowly diving under another and entering the Earth's deeper layers. In this case the role of environmental protection groups and the use of scientific evidence in disputes became the topic.

The weather is, as you might expect in the UK, an endless source of questions. Why does air pressure vary in the atmosphere? What causes the humidity? Why are clouds white? A particularly windy day prompted an interesting weather-related discussion described below.

This example illustrates two of the many routes that discussion can take after an everyday observation. In this case, the answer wasn't immediately known, but the question opened up fundamental ideas in physics, relating to everyday observations. Then, in addition, later investigation via the internet led to the discovery of specialised research that answered the question in an easily accessible way — bringing contemporary science closer to home.

These examples, drawn from observations of the world around us, reveal how easily the highly specific contexts of everyday life — a beach, butterflies,

---

From Everyday Events to Fundamental Science

The weather as a starting point

On a day when a strong wind had battered everyone on their way to the discussion group a person was inspired to ask: "How do butterflies manage to steer a course in the wind? Is the wind a solid thing or full of eddies?" Despite the potential for a zoological exploration about the life of butterflies, the thread on this occasion led into the basics of fluid flow and, unexpectedly, into a piece of contemporary research. Discussion opened up with the issue of flight — Leonardo's sketches of birds and flying machines — then moved into an explanation of smooth and turbulent flow in rivers and air streams. People remembered the eddies they had noticed looking down from bridges at water flowing past a pier. The cause of the spinning water was quickly grasped as a frictional effect: the result of fast-flowing water having to pass by slower water held back by the solid pier. After this excursion into fluid dynamics a quick internet search revealed a study at the Rothamstead Research laboratories near London had indeed been tracking butterfly flight paths across the landscape, using radar. So, a direct answer to the original question was soon found, to complement the broader foray into fluid flow: "Butterflies recognise and respond to landscape features ... they are capable of similar levels of straightness during flight, regardless of prevailing wind direction."

a sunset — can be the point of entry to profound insights into the fundamentals of science. Sadly, in science education the sequence is so often reversed. General theoretical propositions are put to students, whether or not they find them relevant, which then have to be "learned", often by rote. Applications of these propositions in real-life situations may or may not be added as an afterthought. A systematic review of context-based approaches in science education[viii] concluded that approaches that use the contexts and applications of science as the *starting point* for the development of scientific ideas foster more positive attitudes to science generally without loss of understanding.

It is also clear from these examples how easily discussion led by the interests of ordinary members of the public, as opposed to scientists or science teachers, veers off conventional scientific paths. In a permissive environment, personal experiences are exchanged, thoughts are bandied about freely and misunderstandings exposed without embarrassment. Yet, as the illustrations show, the motivation generated by such grounded discussion creates a thirst for explanation, leading on to deeper scientific enquiry and often, a return, after unpredictable excursions, to the original scientific point.

Health

A second rich source of questions is the area of health, often related to the concerns people have about their children and ageing relatives. A person in one discussion group had been reading a book by the psychotherapist Sue Gerhardt which showed, (in the group member's own words) that "relationships can make the brain grow". She had seen that MRI scans can reveal that a baby has been neglected. She asked whether parents had always been affectionate to babies and children. A flurry of questions followed: How do chemicals alter behaviour? Do hormones affect how the brain thinks? Why are teenagers grumpy — is there a physical basis for the "Doh!" syndrome? What makes us have ideas?

This illustrates just how wide-ranging the questions can be once a significant issue has been raised. Of course, to address them all in one session would be quite impossible and indeed this is one reason why, in traditional syllabuses, topics are organised in carefully planned steps. In this alternative approach, however, questions are not answered directly, but instead serve as entry points for learning about fundamental concepts. In this instance, the idea of regulation of hormone levels through feedback loops was introduced and compared to thermostats in the home. More detailed information about the pituitary gland was sought out after the session and used in the following session to explain the link between electrical signals arriving from the brain and the release of hormones into the bloodstream. Eventually the question of teenage grumpiness was addressed through a visit to a clinical endocrinologist at the local hospital. She provided an introduction to hormones — their immense variety of form and their specific role as chemicals that act away from their source — and then addressed the group's real-life questions. The teenage issue was explained as partly due to the rapid increase in sex hormones during puberty and partly by broader psychological factors: the quest for identity and the shift away from parents.

Health issues arise frequently and are often an entry point for discussions on the way the body works. These naturally lead to insights about major bodily systems and organs: nerves, hormones, blood vessels, the brain, glands and so on. But, interestingly, more detailed anatomical issues tend not to be pursued; instead the human body itself often becomes the starting point for deeper discussion about the fundamentals of chemistry and biology.

The question was once put: "Why do our bodies need iron? Is there a parallel between iron in the blood and in the Earth?" This revealed a deeper conceptual issue about the nature of "chemicals". The very word "chemical" is widely perceived as negative — something artificial, man-made, potentially threatening to the balance of nature. The idea that a (pure) chemical is identical whether made in a lab, mined from the Earth or found in the body is difficult to accept and, as with other counter-intuitive concepts, becomes understood gradually as it recurs in various contexts. Typical questions raised in groups about the nature of chemicals include:

> "What is the difference between chemicals and organic remedies?"
> "Why is St John's wort more acceptable than a drug?"
> "What is the difference between a vitamin and a mineral?"

An example of a journey from family health to fundamental chemistry arose in a discussion about hormones. One person said, "Now we know that steroid simply means a type of chemical, it challenges the general perception that steroids are a 'bad thing', to do with doping in athletics." Her question at the next discussion session — "Are all hormones steroids, and all steroids hormones?" — led to the realisation that there are chemical groups (particular arrangements of atoms) and that chemical compounds are generally made up from a limited number of such groups rather than an arbitrary assembly of atoms.

Often, questions that begin with a health issue lead into basic biology. For example, a discussion about the growth of embryos led on to the question "How do cells become different from one another?" This acted as a cue for an investigation into cell differentiation and a visit to a cancer laboratory studying how cells sense their position relative to one another. On a different occasion, the question "How does the body know that puberty is due?" led to an exploration of the time frame for child development in humans and the role that hormones play in it. "Why do we live long after child-bearing age?" proved the starting point for discussion of the mechanisms of evolution — the rate of reproduction and survival of those best adapted to the environment. It also led to more recent social research about the role of grandparents and other members of society in the upbringing of infants. This interesting new angle, close to the hearts of several participants, led to discussions that criss-crossed between the social aspects of kinship and evolutionary theory.

The brain is a frequent source of questions in discussion groups:

> "How do we associate meaning with words and sentences?"
> "Do hormones affect how the brain thinks?"
> "Could MRI scans test whether alternative therapies work?"

The internet proves particularly helpful for these kinds of enquiry, providing up-to-date diagrams of the areas of the brain and helpful three-dimensional representation using animation. Discussion of the geography of the brain often leads to more detailed discussion about neurons and the mechanism of nerve transmission. Fascination with how signals transfer from one neuron to the next at a synapse lead into a wide range of issues of wider social significance. For example: deficiencies in neurotransmitters may lead into discussion of the physiological basis of mental disorders, such as depression; insight into the activation and inhibition of receptors can easily lead into discussion of the mechanisms of addiction and anaesthesia. One discussion on neurotransmitters started from the interesting question: "Where are all the adrenalin molecules *before* you get to fight or flight?"

Of particular significance, given that discussions are free to range across traditional subject boundaries, are questions that probe the connection between the animate and inanimate:

> "Is the electricity in nerve cells the same as in a car battery?"
> "If you assembled the right chemicals and electrical impulses would they give rise to thoughts?"
> "What is the connection between proteins, which I know are essential for me to eat, and molecules?"

The idea that sodium ions, electrons, amino acids, fats and minerals and all the other entities of chemistry and physics are identically the same, whether in living or inorganic matter, is strange and unfamiliar to many people. Grasping it through careful discussion in relation to people's lived experience can be highly enlightening. This issue is taken further in Chapter 6.

These examples of discussions inspired by health questions illustrate a key feature of the open-ended approach. Although initial questions may appear to come from an immensely varied range of contexts — embryonic growth, adolescence, grandparenting, steroid drugs, herbal remedies — the path of discussion tends towards a limited number of fundamental concepts, rather than a plethora of unrelated ones. The merit of this, from the point of view of deep learning, is that key topics are visited and revisited many times over. The concepts of molecules, chemical messengers, nerve transmission and brain topography, for example, become gradually more familiar over time and can lead on to deeper scientific enquiries, unrelated to the original specific context.

Domestic life

The family and its health is an important area for questions about science, but other aspects of domestic life also arouse curiosity. Some are practical and appear to demand snappy answers:

> "How does a radio work? It seems like pure magic."
> "What's happening when an egg cooks?"
> "Why doesn't electricity leak out of the sockets?"

However, brief and direct answers such as "eggs turn white and solid because the protein in them is denatured at high temperatures" are simply too dense to be taken on board. Even when they are comprehensible, they tend to close down the possibilities for deeper exploration of fundamental ideas. These answers simply deflect the enquiry by presenting a further set of mystifying concepts, such as the denaturing of proteins. A direct response to the question of how a radio works would mean covering a huge range of complex ideas — wave transmission, oscillating electrical circuits and amplification theory — too much detail for a two-hour session. But such a complete explanation is not really the motive behind the question; the point is to use the question as a starting point to develop a few basic concepts to replace the pre-existing sense of total mystery or magic. So the focus is on ideas: the idea of electromagnetic waves permeating the space around us, the idea of sound waves superimposed on top of these, and of the radio tuning circuit being designed to select out one particular wavelength from the myriad that surround us. These become the substance of the discussion.

Questions sparked off by everyday aspects of home life can easily lead into deep science. "Does salt disappear when it dissolves?" was a simple question that once led into the whole question of the bonds that hold atoms together in a substance; a topic that was revisited later with the question "What causes glue to stick things together?" Sometimes questions originating in the practical world of the kitchen can be surprisingly abstract. "Do detergents occur naturally or were they invented?" and "When you put on weight are you putting on molecules?" are both interesting ways of looking at fundamental issues in chemistry. Some examples inspired by the bathroom are given on the next page.

News and events

From time to time events in the world become the obvious starting point for enquiries — a volcano erupting, the arrival of cloud computing or the discovery of the Higgs boson, for example.

When the former Russian agent Alexander Litvinenko died in London from alpha radiation, questions were raised in one discussion about the nature of nuclear radiation — how far it penetrated, what it did to the human body and so on. Because atoms emitting the radiation had been ingested, the discussion focused on what radioactive material actually consists of, the different kinds of radiation emitted and how the various kinds of radiation interact with molecules in the body, particularly DNA with its

Questions from Domestic Life

Tales from the bathroom

Kitchens and bathrooms inspire questions about science on a regular basis. One discussion began simply enough with a question about why radio stations gradually fade away as you drive from one place to another. The group began thinking through how the waves radiating from the transmitter must spread out over an ever-larger area as they travel away from the source — rather like ripples on a pond. The waves must get weaker as they do so, just as pond ripples eventually fizzle out.

Conversation moved on to the opposite effect — cases where the energy in a wave can become more concentrated. Waves can be deflected from their straight-line paths by means of prisms and mirrors and lenses. Suddenly, a person in the group recalled a domestic emergency in which a burning smell wafting down the corridor was traced to a towel in the bathroom. It transpired that a shaving mirror had been angled in such a way that sunlight falling on it was bouncing off the curved surfaced and, by great misfortune, focusing just where a towel was lying — a clear example of the way in which waves can be bent so as to concentrate their energy into a small spot.

Talk of bathrooms reminded another person that their gas geyser had a tendency to deliver an electric shock to anyone pointing a finger at it too closely. A foray into static electricity, the flow of current and the role of insulating material followed. On another occasion, discussing how your senses can deceive, a person remembered how easily your body can "sometimes get hot and cold confused — you can feel a cold tap on a bath as hot". The distinction between actual temperature and the bodily sensation of "hot" and "cold" became the topic of discussion.

potential for damaging future generations. In effect the complete syllabus for nuclear radiation at school level was addressed in an hour or two because the relevance of it in current affairs was so stark.

A less sombre instance was the discovery of the Higgs boson in 2013. This event, featured prominently in the news, appealed to a sense of national pride; unusually, everyone seemed to be talking about a scientific story. It provided an opportunity to discuss aspects of physics normally reserved for undergraduate or even postgraduate courses. The basis of the so-called "stand-ard model", that describes the various particles considered to be the ultimate constituents of matter, was explained. Because of the topicality of the story, well-written blogs were immediately accessible, and so key questions about the nature and role of the Higgs boson could be answered on the spot using a smartphone.

These examples illustrate yet another feature of the open-ended appro-ach — the power of relevance. Issues that have an obvious relevance — whether practical (as in eggs being cooked) or abstract (as in bosons being discovered) — are highly motivating. Topics that might have seemed turgid at school suddenly become alive; questions flow, ears are opened and efforts made to struggle with unfamiliar ideas. Of course, by responding to topical issues, the curriculum that emerges over time is something of a patchwork, but the gain in attention and persistence is enormous.

The universe

Curiosity about scientific ideas is not restricted to the immediate and practical world — our environment, our health, our domestic life. People are equally fascinated by the more remote context of our lives: the universe in which our planet exists, its origin and destiny. Some enquiries start with thoughts about the physical processes at work outside our own planet. Others start from news stories about potential threats to life and others simply pursue people's curi-osity about our ultimate surroundings.

As the nature of gravity is gradually understood here on Earth, thanks to discussion about tides for example, questions arise about how it differs else-where in the universe. Weightlessness in spacecraft is now regularly seen on TV programmes but trying to work out what causes it or how low gravity affects motion on the Moon or what the complete absence of gravity in outer space means is no easy task. "If a ball were dropped in zero gravity does it bounce back to the same level?" asked one person (before registering that there's no "dropping" without gravity). "Why does a rocket have to have a large g-force to leave the Earth, can't it just go at a slow, steady speed?" asked another. Trying to answer these questions compels people to think afresh about what is meant by force, acceleration and steady motion in their daily lives — in cars and planes, lifts and cranes. In effect, Newton's fundamental laws of motion, rarely remembered from schooldays, gradually begin to make sense. The idea that a spacecraft will simply continue moving as it is, in a straight line forever, without a motor, becomes more plausible when a long journey to Mars is actually happening.

The sheer emptiness of space is a source of repeated fascination. Is there really nothing there? In what she described as an "epiphany moment" one person suddenly realised in amazement that whenever we look at a distant star, the light from it has encountered nothing at all in its years of travel up until the moment it hit her eyeball. Space had been entirely empty throughout its journey — had it not been so, she would not have seen it, of course. In another discussion about starlight a person asked "Why are we not dazzled by starlight at night? After all if stars are so hot and bright — burning fiery furnaces — why don't they dazzle us like the Sun?" This opening led into insights about the way waves travel outwards in ever-expanding spheres, spreading their original energy ever more thinly, appearing ever dimmer. In effect the so-called inverse square law had been revealed and understood, through reasoning and discussion alone.

Apart from its emptiness, space also leads to queries about the meaning of hot and cold. As one person put it in discussion: "Wouldn't a spaceship get hot near the Sun even though space is cold?" The absence of matter in space forces people to rethink the idea of warmth, associated as it is so closely with the air here on Earth. The concept of radiated heat travelling through the emptiness of space, as opposed to heat that is conducted or convected in the presence of matter, becomes clearer as the passage of the Sun's energy through the vacuum is grasped.

Anxiety about space may also be the trigger for discussion. News of a meteorite on an Earth-bound trajectory or gusts of solar wind threatening to overpower Earth's magnetic defences open up enquiries about the Earth and its neighbours. One such discussion, beginning with "What are cosmic rays? Are they radioactive? Do they just go through you?", led into a comparison of the various forms of radiation and their effects on the human body. Again, the freedom to move across conventional topic boundaries enabled this discussion to flow seamlessly from the nature of the stream of particles emanating from the Sun (the solar wind) to the nature of radioactive radiation (alpha, beta and gamma types) and on to the effects of magnetic fields on electrically charged particles. So, what had started out as an anxious response to a newspaper scare story moved on to an exploration of how the Earth's magnetic field diverts potentially harmful streams of solar particles and led ultimately to insights into fundamental aspects of physics.

Interest in the origin and early life of the universe is a constant source of fascination. People are curious about where the matter that makes up our bodies, our environment, the planets and stars, originally comes from. A questioner once asked: "Is there more matter in the universe than was created in the Big Bang? Are we just recycling the original atoms?" More generally, the question "What is a universe?" was asked as theories of multiple universes began to appear in the popular press.

These examples illustrate how extending thinking beyond the bounds of our Earth can help clarify fundamental concepts. Almost all our intuitive grasp of natural processes derives from our personal experience on Earth. Thus from experience, moving objects slow down if they are not pushed, light dims as it passes through the atmosphere, warmth spreads around us. Imagining processes as they occur in space, not only fuels our sense of wonder but also helps us distinguish parochial behaviours here on Earth, with its particular atmosphere and frictional and gravitational forces, from those in the vast regions of outer space in which these are barely perceptible.

Deeper scientific questions

As some of the vignettes given above suggest, the path of an open-ended discussion easily strays far from the everyday setting in which it may have begun. Kitchen matters lead into deep chemistry, family health into molecular structure, and holiday observations into the joys of electromagnetic theory. From initial questions based on common experience a stream of secondary questions soon arise, which often reflect deepening curiosity about the scientific concepts that have been introduced. In the next chapter we look in detail at some of the fundamental themes that characterise discussions. Here we illustrate some ways in which deeper enquiries can be sparked off by initial questions from everyday experience.

Many questions probe the ultimate nature of matter, but often start from something familiar, as the example on the next page illustrates.

Clarifying the distinction between words like "element" that refer to substances and words like "atom" that refer to structures, may lead on to

From the Familiar to the Fundamental

Atoms and elements

One particularly interesting discussion began after a group had visited an open day at a local university. Having witnessed a demonstration of three-dimensional (3D) printing, one group member said she "had the idea that an atom is not the same as an element — an element seems like something you can get hold of". Another member ventured, "What do we mean by a compound thing like, say, a daisy — it has petals, flowers, stalk, etc. The daisy eventually dies but don't the atoms live on? So they can't be daisy atoms." "The atoms are recycled," thought another group member, "aren't the main atoms in living things oxygen, carbon, hydrogen and nitrogen? When a daisy dies microorganisms digest it. The atoms are recycled and will reappear in other plants later." Ultimately, the existential question was put: "Do the atoms live on when a person dies?" In essence, this discussion was an exploration of the meanings of the words "element", "atom" and "molecule".

A subsequent questioner asked in the same discussion: "Is anything made solely of itself — gold, for instance?" This brought out an intuitive sense of what an element is — something pure, not made up from other substances — and opened up a discussion of impurities and the compound form in which substances tend to occur naturally.

questions about the ultimate building blocks of matter. The common question "What is the smallest thing you can get down to?" opens up discussion of the menagerie of fundamental particles, some of which people may have heard of, most of which are unfamiliar.

The concept of a molecule as a group of atoms bonded together gives rise to many questions about their size, structure, role and so on. An interesting question someone once posed — "Can a molecule be made up of other molecules?" — became the starting point for a stream of further questions and discussion:

> "If you break down a molecule are all the bits also molecules?"
> "What is a bond ... is it part of the molecule?"
> "What is in between the H_2O molecules in ice?"

"Are two magnets sticking together a chemical bond?"
"Why don't water and oil molecules mix?"

After an exhausting session tackling all these questions one participant sat back incredulous, exclaiming: "Molecules, atoms, bonds, valence electrons — it's all so incredible, did someone make it up?"

Another area in which deeper exploration of scientific ideas often takes off is genetics, as the illustration below shows.

From Everyday Observation to Deep Theory

Bluebells and DNA

A few white bluebells had been noticed one springtime, in amongst a carpet of blue. About one in a hundred seemed to be white and were spread randomly amongst the blue. This reminded someone in the discussion group to recall something they had read about the Tureg people in Tunisia who apparently have blue eyes and dark skin. Another person had read a newspaper article claiming that redheaded people will disappear one day because there was not enough marriage between redheads. This gentle exchange of observations led a member of the group to ask: "Does albinism get passed on? If it did, wouldn't there be more albino individuals? Why is it so rare?" Others added to the point "What about eye colour? Different versions remain in the population ..."

These questions enabled different biological processes to be explored. The idea of an unusual gene being inherited from both parents was introduced in explaining albinism. Apparently, the effect is to damage the body's production process for the skin pigment melanin. But as someone pointed out, "if melanin helps protect the skin from harmful ultraviolet radiation it seems odd that the defective gene persists in the population, generation after generation". The general point was made that disadvantageous traits can persist in a population if the genetic alteration also confers some other greater advantage to the individual.

Discussion moved on to more fundamental questions:

"How do changes arise in genes?"
"Is the genetic mixture in a fertilised egg exactly half from each parent?"

(Continued)

(*Continued*)

The concept of genes as sections of the very long molecule DNA was introduced. The information coded by a sequence of chemical groups in the DNA was described and the mechanism by which body components are created set out. A gene usually provides the information needed to produce a protein for the body; these then become the building blocks in the body and the chemicals (enzymes) that make things happen. Defective genes (or mutations) may be unable to produce a necessary protein or produce a protein that doesn't work. Occasionally, however, a gene mutation can produce an altered protein that actually works better — the raw material for evolution.

A few simple observations opened up a series of fundamental concepts in reproduction, inheritance and evolution — material enough for months of further discussion.

As the examples show, issues in everyday life provide a boundless source of initial questions about science, and the same sense of curiosity also attaches to the scientific ideas subsequently discussed. The nature of the elements, the structure of DNA, the laws of inheritance — these become the contexts for deeper probing. In this sense, the question of relevance, the closeness of scientific topics to aspects of everyday experience, seems to be important primarily at the entry points into science. Once initial obstacles have been overcome and scientific insights begin to develop, fascination with the ideas themselves seems to take over. Questions follow to clarify understanding and move it to the next level. Bizarrely, gaining insight into a scientific concept very rarely seems to satisfy the questioner. As soon as an idea is grasped a kind of irritation seems to set in, urging the enquirer to move on, to confront the next mystery immediately! In this respect, the curiosity that drives an ordinary citizen into science can be quite exhausting — and closer than one might think to that which drives professional scientists!

Beyond the natural sciences

As you might expect with enquiries that start from everyday life, the path of discussion often moves beyond the bounds of the natural sciences. The links people make with scientific ideas often lie in the social sciences, arts or humanities. It is an unfortunate fact of our times that the natural and human sciences have drawn apart with the increasing specialisation needed to keep abreast of the expanding body of knowledge. Fewer and fewer people are able to follow developments in all spheres and the result is what is now seen as two cultures, as the writer C.P. Snow captured it in his 1959 Rede Lecture. It is now hard to imagine the possibility of a culture in which an educated person could equally embrace the natural, social and human sciences.

The problems that humanity faces, however, are affected neither by the limitations of man's understanding nor by the way the divisions of knowledge are defined. In the contemporary world they seem to derive increasingly from the human and social condition as much as from the material world. Climate change not only requires all the understanding we can muster in atmospheric science, oceanography and power generation but also all we can foresee about its implications for peoples and nations through the geographical, sociological, political and economic sciences. In a similar way, problems in the field of public health, such as the implications of genetic research, will call for expertise across the board, from ethicists, lawyers, politicians and insurers as well as geneticists, microbiologists, epidemiologists and clinicians.

With this background in mind it is not just interesting that free-running science discussions so readily move beyond traditional scientific domains, but it's also heartening. Links made between public affairs and scientific ways of thinking are to be celebrated as an important aspect of citizenship. Discussing biological questions, for example about the brain or hormones, can be the cue for people to exchange experience and thoughts about emotion, child rearing and psychological or complementary therapies.

Discussion frequently moves from a scientific starting point outwards to wider psychological and family health issues and ultimately back to some scientific learning point. For example, in a discussion triggered off by a TV programme on conception and embryo development, the idea was introduced that an embryo competes with the mother for resources, for instance by depleting her bones of calcium. Thinking about how some babies are wanted and others not, one person asked, "How do emotions influence the embryo's growth? How do they affect the biochemical signals? Does the infant's early greed later become manifest as greed for money, love and fame?" An idea from the psychoanalyst Melanie Klein was introduced, that if a baby is not being fed, it believes someone else must be getting the food. Questions were asked about "when a baby's sense of identity arises — the sense that this is me, that is you". A radio programme was recalled about pregnant mothers who diet setting the embryo's metabolic rate low, leading to babies that eat more than needed and becoming obese. Ultimately this discussion settled around the use of drugs in psychological therapies and the way in which drug molecules interact with bodily systems.

Another discussion began after someone had read about experiments in social psychology that showed certain emotional reactions occur repeatedly in all cultures. There is now a concept of universal types of emotion: disgust, affection, anger, fear and so on. Initial discussion focused on whether emotional triggers differ according to social and cultural context. The post-modernist and positivist viewpoints were contrasted and ultimately seen as not necessarily so far apart. As one person put it "post-modernism tries to pull the rug out from under things, to reveal things that are hidden, like institutional racism". "Perhaps post-modernist ideas like this have just come to be accepted now — the existence of latent tendencies," said another. This prompted a further social observation: "The pace of cultural change is now so fast: only a few decades ago the MCC [Marylebone Cricket Club] accepted South Africa's request to deselect cricketer Basil D'Oliveira; today children in poor parts of rural China are playing computer games." Ultimately this wide-ranging discussion about social change circled back to scientific issues through the example of Galileo's experience of resistance to new ideas by the church in Rome.

Sociological issues can equally well be the starting point for scientific discussion. One such began with a visit to a sociologist studying the educational achievement of women and their roles in the workforce. He pointed out that females now outperform males in almost every level of education and in every country in the OECD (Organisation for Economic Cooperation and Development), yet are not appearing proportionately in top employment positions. Discussion of the possible causes of this centred on the interplay of neurological and psychological factors with the more obvious economic and cultural ones.

Another discussion was sparked off by a study of the history and causes of violence: Steven Pinker's *The Better Angels of our Nature*. Research evidence in the book shows a secular downward trend in violence, both in war and crime and in the home. Discussion of this scientific evidence elicited contributions based on participants' personal experience and on their worldviews. Some favoured sociological explanation in terms of the distribution of power, others referred to their experiences of bringing up children. "These days we tend to talk to kids more and give them accurate information," said one. "Kids are influenced by their parents, their peers and their biology," said another. A third raised the question "Do parents transmit values?" Debate returned to the natural sciences through the issue of childhood anxiety, attachment to the mother and the role of hormones, particularly cortisol, in mediating anxiety — "Is the benchmark level for cortisol (a kind of thermostat for anxiety) inherited? Is it influenced by life in the womb?"

A more detailed illustration of the way in which perspectives from the natural and social sciences interplay in a particular case is given below.

The Interplay of Natural and Social Science

A discussion about epigenetics

A Channel 4 documentary on epigenetics — the passing on of characteristics acquired from the environment during an individual's life — sparked off a discussion lasting several months. The documentary had described the case of pregnant women in Amsterdam during the Second World War who had suffered malnutrition and unsurprisingly had smaller than average babies. However, when these babies grew up, their children were also smaller, although their parents had not been malnourished. This suggests the tendency had been passed on to the grandchildren epigenetically.

This recent discovery appeared to contradict the basic tenet of Darwin's theory of evolution: that characteristics cannot be acquired from the environment and passed on through the genes. Questioning led into Darwin's theory and the challenges to it from Lamarck. Further queries led into the nature of genes, as we understand them today: what do chromosomes do? What is a recessive gene? From this genes were discussed as defined lengths of the DNA double helix molecule. A deeper question was asked: how do we know where one gene ends and the next begins on a length of DNA? From this a detailed description of the DNA molecule was entered into, including the various chemical groups that make up the molecule and the

(*Continued*)

(*Continued*)

distinctive four-letter code based on four specific chemical groups. The recent discovery that other chemical groups (methyl groups) can be added to the DNA molecule under particular environmental conditions — the epigenetic mechanism — moved the discussion into contemporary issues in genetics, both scientific and social.

The internet, with its graphical images, proved a vital source for information about recent studies. It showed how, for DNA to do its job in making proteins, the spirals into which it is wound need to be uncurled and the genetic code "read off". The amount of the very long DNA thread that is uncoiled seems to be determined by the presence of these "methyl groups" attached to the DNA thread.

Discussion moved on to the social implications of these discoveries, when it was found that research shows that the density of methyl groups ("methylation") is affected by stress and other environmental factors. "What is stress?" asked one person. "Is methylation all bad?" asked another. "Does the extent of methylation vary throughout life?" asked a third. "If stress is related to socio-economic status, what are the social implications of epigenetics?"

Sometimes the political aspects of scientific knowledge become the issue. In one example, a scientific discussion about the way in which drugs interact with the body led on to political questions about the classification of illegal drugs. A clash between a scientific adviser and the secretary of state had featured in the news. Their public dispute over declassifying cannabis had pointed out the limitations in the role that scientific evidence plays in controversial areas of public policy when the views of the electorate are paramount. A subsequent visit to the discussion group by a government scientist provided insight into the part science plays in political decision-making. This discussion provided insight into the importance of both the social and physical sciences in the practical world of policymaking on matters as diverse as airport queuing, DNA databases and experimentation on animals. As the government scientist put it:

Scientific research is not always conclusive, and you have to use it to inform rather than dictate a conclusion, especially since there are other factors, for example political judgements, that have to be taken into consideration. Longitudinal research is expensive, and with many groups you can't do randomised field trials — prisoners, for example. You often don't know *why* things occur, but that's not always necessary. After all, John Snow worked out the cause of cholera in Soho by studying the geography of the epidemic without knowing anything about microbes.

Sometimes enquiries, particularly in areas of physics, tip over into quasi-philosophical discussion. In one example, discussion centred on the recently opened Large Hadron Collider (LHC) at the European Organization for Nuclear Research (CERN) in Geneva. An explanation was given about the proton beam, how it had to be created, channelled extremely finely, bent in a continuous circle and accelerated before hitting its target. Following the question "Does the beam have to be protons?" an explanation was given that in order to bend a stream of particles, they have to be charged. Unexpectedly the next question was simply "What is charge?" — an apparently simple question. The questioner went on to explain: "I have never understood what charge is, though I have always found people, especially men, seem happy to use the word constantly; yet they don't seem able to explain what it is — they just get on with using it."

"What is charge?" — a profound question that is simply side-stepped in school and university education by describing it as a quality that a material can have which causes repulsion or attraction when two types come close to one another — like a comb and hair or a rubbed balloon and nylon jumper. The concept is rapidly captured in mathematical form and given the symbol Q. The original qualitative question is gradually forgotten. In the more demanding atmosphere of an adult discussion group this level of response is rarely acceptable. In this case discussion moved on to a philosophical plane, focusing on the nature of explanation itself: the problem that the fundamental qualities in the material world, such as "charge", cannot really be explained in terms that are more fundamental still. Charge is not a quality that can be seen, or touched or sensed directly. It is an intellectual concept invented to explain an observed process — attraction and repulsion between objects. It provides a model of behaviour from which theory can be developed which proves powerful in predicting and explaining countless other behaviours. On this occasion, discussion of this somewhat chilling truth led the group to realise how much we rely on analogy to help us understand something new in terms of something we are already familiar with. Electric current flows: it's like a river; massive objects attract: they are like magnets. But at the fundamental level, charge and mass just are; we invent these qualities and give them names in order to understand and predict the world around us.

A more detailed example of an even deeper philosophical discussion about whether the colour red actually exists, given the wave nature of light, can be found below.

The Interplay of Science and Philosophy

Does colour exist?

A discussion once began with a person asking why the speed of light features so prominently in Einstein's theories of relativity. She went on to ask: "Isn't light the only thing like itself? It has no mass. Aren't all other things made of particles?" This led to an explanation of light as a wave, consisting of tiny electrical and magnetic disturbances oscillating in strength extremely rapidly. This led to the seemingly innocent question: "Can you see light waves?"

The paradoxical answer is that you can't, unless they happen to enter your eye! For example, in outer space you wouldn't see a torch beam from the side. We only see light from a torch beam when it is reflected off particles in the atmosphere into our eyes; in outer space there are no particles (almost). So "seeing" is entirely dependent on the presence of eyes, more precisely on the interaction of light waves with molecules in the retina of the eye. With this unsettling thought in mind one of the group said, "If colour is about us seeing, then that jumper is not red if no one is looking at it!"

An explanation was given that the molecules in the jumper absorb most of the different wavelengths from the spectrum of sunlight falling on them but reflect just a few wavelengths — red wavelengths in particular. This is what we actually see as the jumper. Dyes are simply molecules whose atomic structure means they absorb most colours, but reflect some: in this case red. They are added to the material of the jumper. So in a dark room, the molecules are still the same in the red jumper, but they are not activated because no light is falling on them. So, "Red light and a red jumper are different kinds of thing then," commented one of the group. The group came to the conclusion that there is nothing "lighty" about a light wave itself — it only differs from an infrared wave or a microwave in the length of its wavelength. It is the way the eye responds when a wave impinges on the retina and the way the brain interprets this that constitute the human sensation of light.

Another popular area of discussion draws on linguistics and child development, as well as neuroscience. The ways in which our brains process language and our children acquire language are topics close to everyday experience, as well as being intrinsically fascinating. A discussion on this theme extended from neuroscience to cultural comparison. Triggered off by a member of the group who had read the book *Musicophilia* by Oliver Sacks, discussion turned to the way pathways are sculpted in the brain through repeated experience. The question was asked whether this could happen later in life, by learning the piano at age fifty, for example. Questioning moved on to language development:

> "Does the direction people read (left to right as in Europe or right to left as in Israel) affect the brain hemispheres?"
> "Are the same parts of the brain used in sign language as spoken languages?"
> "Are left-handed and right-handed tendencies the same for babies?"
> "How do babies develop language out of the cacophony of sound around them?"
> "Is forgetting (and remembering) the same for the written word as for the spoken?"

Into this welter of questioning people contributed from their own experience and reading. One talked about the Japanese, whom she believed use both hemispheres for language because it is both visual and spoken. Another suggested there must be several areas of the brain associated with language — auditory, visual and a semantic area. The discussion inspired deeper study of the areas of the brain using the internet after the session. The questions went further at the subsequent session: "Would people with damage to the Wernicke's area fail to recognise nonsense" unlike a stroke victim someone knew who had "become frustrated with their own nonsense"? Discussion flowed on to the role of metaphor in the evolution of languages, the phonological origin of early words "mama, papa, baba" and on to forgetting ("Is it nouns rather than verbs that go first?") returning finally to the common experience many now have of relatives living with Alzheimer's.

Taken together, these examples demonstrate something of the range of disciplines surrounding the natural sciences that are entered into when the path of discussion follows the line of curiosity. We have seen how neuroscience may link to experience of therapies; optics to the philosophy of colour;

socio-economic disadvantage to brain physiology and epigenetics. Discussions driven by live issues in the personal experiences of the participants can range even more widely: sometime into education — the processes of learning; sometimes into religion — ultimate causes beyond experimental explanation; and sometimes into history, geography, anthropology, cultural studies and linguistics.

With such scope for discussion and such freedom to roam, what chance is there that the scientific ideas will hold their own against the lure of other, perhaps more scintillating ideas in other subject areas? This is indeed a major challenge to such a liberal approach to science learning. The evidence of hundreds of sessions in which this liberty has been granted is that science holds its own very well indeed. Threads of discussion that move into the social or humanistic spheres regularly return to their scientific root. The interplay between the discomfort of unfamiliar scientific concepts and the certainty of personal experience enable difficult questions to be posed, exploration to proceed fearlessly and, as a consequence, challenging concepts to be addressed with renewed confidence and vigour.

The importance of interaction between the natural, social and human sciences can hardly be overstated. The historical separation between these disciplines and the cultural distance that has developed between their communities mean this all too rarely occurs. The word "consilience" has been adopted by the eminent American biologist E.O. Wilson to describe the rapprochement needed between the spheres. In his book of that name[ix] he points out that "the ongoing fragmentation of knowledge ... is not a reflection of the real world but an artefact of scholarship" and goes on to assert that "when we have unified enough certain knowledge, we will understand who we are and why we are here". Citizens with the breadth of their experiences in the world have as big a part to play in this process as scholars and professionals with their specialist contributions.

Dampening Curiosity

We have focused thus far on what excites our curiosity and where the path of discussion leads once it has been aroused. It draws heavily on positive experiences in a range of facilitated discussion groups. However, the conditions that encourage people to pursue their natural curiosity seriously, to venture their untutored thoughts about science in the presence of others, are not easy to find. In many areas of life, we are inclined to leave our partly formed thoughts undisturbed; our sense of personal failing and fear of humiliation serve to keep a firm lid on them. For many adults, science (and equally mathematics) is sensed as a weakness in their overall grip on things; unhappy experiences in trying to grapple with it earlier in life may have left its mark. Given this degree of vulnerability, encouraging people to expose their scientific thoughts can almost be a kind of therapeutic process. Success in opening up science for adults, as for other areas of vulnerability, depends on careful and skilled facilitation, mutual support, congenial surroundings and a willingness to participate. In this section we look at the factors that dissuade people from pursuing their scientific questions, and what it is that so easily dampens their natural curiosity.

There seem to be many reasons why people cease pursing questions and paradoxically, one of them is because too often they are just given direct answers! At first sight, answers seem to be just what are demanded: preferably something immediate, understandable and concise. On the whole this is what people around us, and indeed many websites, are only too keen to offer.

The problem is that a closed answer to a question is not necessarily what people really want. Sometimes such an answer may fix an immediate practical problem — which is the live wire in an electric plug, for example — but does it satisfy curiosity? In practice a question about science can act as the herald of a rich and unpredictable flow of discussion, just as questions in literature or history are expected to do. After all, the question "Does Hamlet's speech to Yorick's skull represent a philosophy of death?" would hardly be satisfied by the answer "Yes". Beneath a scientific question may well lie an equal expectation that complex ideas are to be explored from many points of view and something deeper learned.

An interesting example that occurred in a discussion group is given on the next page.

So any influence in a group overeager to supply the narrow answer can be the kiss of death to enquiry. In this respect the female propensity to entertain doubt, share tentative thoughts and hold back from asserting factual answers to open-ended questions, serves the process well. Indeed, are didactics that favour clear-cut answers to factual questions one of the off-putting aspects of science for many women? Certainly the open-ended process described here has proved particularly attractive to them.

Not Answering the Question

Why are clouds white?

A discussion was once started by a person recently returned from two weeks on a Mediterranean beach. "Why are clouds white?" she asked simply. The answer, given crisply on the internet, is "because they reflect the sunlight". End of story. But below the surface of the question lies a rich seam of unexplored ideas. What in fact is a cloud? What indeed do we mean by white? How does reflection work? In this instance, by leaving the question unanswered in the immediate sense, months of discussion followed on the nature of the water droplets and ice particles that make up clouds, the manner in which light waves are reflected or transmitted at the curved surfaces of droplets, the condition under which sunlight is broken up into its component colours, and so on. The discussion deepened with arguments over whether a cloud has an edge, and if not, at what point an entity that gradually thins out can be said to terminate. More advanced physics was probed when it was realised that when sunlight is reflected from the atmosphere rather than a cloud it is not in fact white, but blue (i.e. the sky). Deeper enquiries from a cloud expert revealed that light is scattered differently according to the size and density of the scattering particles. A simple experiment with a drop of milk in a glass of water, demonstrated this neatly, when a blueish tinge was clearly observed through the diluted milk. A question not answered acted as the starting point for an unpredictable journey through the make-up of clouds, components of sunlight and physics of scattering.

Fruitful enquiries can also fizzle out when too many plain facts are loaded into discussion. One person or another may happen to know more about the size or shape or longevity or weight of something and take the opportunity to let others know it. Again, interesting though factual detail may be at some point in a discussion, it all too easily intimidates at the early stages, leaving less-informed participants to withdraw interest, leaving their potentially divergent questions unasked.

Perhaps the greatest barrier to open-ended enquiry is the scarcity of opportunities for interactive discussion about science. For many adults the experience of science in education rarely extended to open-ended discussion.

For most the norm was of science as a school subject, dominated by rote learning, textbooks and exam papers. The idea of discussing science socially, on terms you understood and owned, was inconceivable. For other areas of learning open discussion is welcomed — book clubs flourish, in which individuals share in a free-flowing, egalitarian way; political associations enable something similar for current affairs. Encounters with science more frequently follow the expert-novice norm. Popular lectures by academics and engagement activities organised by interest groups tend to focus on someone else's choice of topic, conducted in someone else's way. Fortunately, recent years have seen some positive developments: cafés scientifiques and other informal opportunities are gradually taking off, at least in some metropolitan areas.

For the determined enquirer, the lack of a discussion group locally may not be a fatal blow. After all, there is a host of popular science books now available everywhere, on a wide range of subjects. But once again, curiosity is easily dampened down if the writing style and use of language fail the reader. As you might expect, popular science writing often attracts scientists as authors. Expert though they may be and celebrated though they may have become, all too often their explanation of scientific concepts leaves the reader gradually losing faith after the seductive simplicity of the introductory chapters. Explaining complex ideas in a narrative that is readable as well as understandable is no mean task for amateur communicators. The net effect of failing to grasp the insight promised may result in the one-time enthusiastic reader ending up less inclined to scientific enquiry than they were initially. Fortunately, the rise of professional writers and documentary makers in science is beginning to make a difference as books and broadcasts are now emerging that manage to engage and entertain as well as inform. The challenge for the potential reader is to know how to distinguish those that will engage them from these that will entrench their fearful feelings.

Fostering Curiosity

For the ordinary citizen, keen to develop an interest in scientific ideas, the key is to find a way to share their curiosity with others. Simply exchanging thoughts, questions and prejudices can activate thinking and inspire investigation. Discovering others with a similar sense of frustration with the lack of scientific background can release the energy to do something about it.

Strengthened with a little social support, ways of getting to grips with scientific ideas might begin to appear. Television documentaries now provide a starting point for "post-match analysis" with friends. Book sharing can help you navigate the choppy waters of popular science writing, helping you identify those that are readable by the typical layperson. The popularity of the various media was measured in a survey for the EU's Eurobarometer, which showed television as the main source of information about developments in science and technology, being used by 65% of respondents. Newspapers and websites came a long way behind at 33% and 32%, respectively, followed by radio at 17% and books at only 14%.

Beyond the passive forms of reading and viewing, a further world of trips and visits becomes a possibility. Science museums are a feature of many metropolitan centres and more specialised venues for astronomy, manufacturing, mining, shipbuilding, health promotion and sports may offer insights into aspects of science through displays or talks with specialists. Locally, clubs and associations exist for enthusiasts in astronomy, geology or wildlife, for example, and often these are welcoming to newcomers, novice or otherwise. On the national scale, fairs and festivals are on the increase, providing demonstrations, exhibitions, discussions and lectures aimed at everyone — a development of the long-standing tradition of making science exciting for children. A more detailed look at these and other practical ways of getting to grips with science is given in Chapter 8.

The other player in this game is of course the science tutor, teacher or discussion facilitator with whom adults curious about science may come into contact. What are they able to do to support the tentative enquirer? What do they need to do to break from traditional approaches to science teaching and to encourage open-ended discussion led by questions from participants? For a teacher accustomed to a well-defined curriculum, abandoning the syllabus, straying outside a subject discipline and dealing with questions for which you don't know the answer can be profoundly unsettling. Science teachers are generally taken to be knowledge experts rather than facilitators of independent learning; their credibility, to themselves as well as others, depends on being well informed and being right. These strictures tend to push a science teacher to be careful to stick tightly to their subject specialism and to control discussion carefully to avoid straying into areas outside it.

To operate in a more liberal discussion format, the tutor needs to swallow hard and welcome the topics as laid down by the participants, regardless of how they fall in relation to her or his expertise. The key is to become a careful listener, to try to understand what lies behind the presenting question. This alone encourages people to develop their curiosity. A further challenge is avoiding answering questions directly but to float them for others in the group to pick up on. Forbearance is essential as anecdotes are offered and distant conversations recalled some of which will seem relevant and others less so. By encouraging a variety of responses a picture begins to emerge of what is known, what is misunderstood and crucially, which direction the scientific input needs to take. In this inclusive atmosphere a major challenge for a teacher is to avoid the temptation to correct "wrong" statements at too early a stage. As for a doctor, the task is to establish the client's authentic view of things first and to sort out afterwards how to approach the treatment. So-called wrong statements are the key to revealing misunderstandings and resistances to learning.

A further challenge for the tutor is to work out rapidly how to respond to topics outside their own area of expertise. Despite the huge anxiety that science teachers tend to have about this problem, it turns out to be much more straightforward than feared. In practice adult learners are only too aware of the bounds of a person's fingertip knowledge. Though there may be the occasional humorous joshing over something the teacher doesn't know, people are well aware that detailed knowledge on almost any topic can easily be found in books and on the internet. The task for the tutor is to look for the underlying scientific concepts beneath the presenting question and to steer the discussion towards them. Thus a question about Alzheimer's may be the cue for exploration of nerve cells or neurotransmitters, and a question on the weather may be the opening for the nature of gases or transfer of heat energy. The choice for the tutor becomes whether to launch into an explanation of the concept directly or to defer it to a later session, giving time for books or websites to be checked out.

For me, the remarkable discovery in working with discussion groups over many years is that, once the game of expert and novice has been abandoned, people are unexpectedly content to let issues hang. Full answers do not have to be given directly or immediately; discussion can continue and detailed knowledge deferred to a later session. The advances in communications tech-nology in recent years mean full command of factual knowledge has become even less of an issue. With ubiquitous smartphones and tablets, people are only too happy to search for information on the spot as discussion proceeds.

So, in conclusion, curiosity about scientific concepts is easily aroused, if the conditions are right. On the whole, people appear to be full of questions and experiences and only hold back through lack of opportunities to explore them safely. Discussion based on this curiosity takes off and sustains when participants are ready to voice their questions freely and share their thoughts fearlessly and a science tutor is available who is willing to relinquish their traditional roles of knowledge expert and curriculum controller.

Endnotes

i. Gopnik, A. (2009). *The Philosophical Baby*. London: The Bodley Head.

ii. Osborne, J.F., Simon, S. and Collins, S. (2003). 'Attitudes Towards Science: A Review of the Literature and its Implications.' *International Journal of Science Education*, 25(9), 1049–1079.

iii. Reiss, M. (2004). 'Students' attitudes towards science: a long term perspective.' *Canadian Journal of Science, Mathematics and Technology Education*, 4, 97–109.

iv. Osborne, J. and Dillon, J. (2008). 'Science education in Europe: critical reflec-tions. A report to the Nuffield Foundation.' Nuffield Foundation, London, p. 15. Available at: http://www.nuffieldfoundation.org/science-education-europe.

v. Schreiner, C. and Sjøberg, S. (2010). *The ROSE Project: An Overview and Key Findings*. Available at: http://roseproject.no/?page_id=39.

vi. National Science Foundation (2004), Public Knowledge About Science & Technology. Available at: http://www.nsf.gov/statistics/seind04/c7/c7s2.htm

vii. European Commission, DG for Research & Innovation (2013). *Special Eurobarometer 401: Responsible Research and Innovation (RRI)*. Available at: http://ec.europa.eu/public_opinion/archives/eb_special_419_400_en.htm#401.

viii. Bennett, J., Hogarth, S. and Lubben, F. (2003). 'A systematic review of the effects of context-based and Science-Technology-Society (STS) approaches in the teach-ing of secondary science', in *Research Evidence in Education Library*, London: EPPI-Centre, Social Science Research Unit, Institute of Education, University of London. Available at: http://eppi.ioe.ac.uk/cms/Default.aspx?tabid=328.

ix. Wilson, E.O. (1998). *Consilience: The Unity of Knowledge*. New York: Knopf.

Chapter 6

Underlying Themes

Introduction

The last chapter gave us a picture of the kinds of questions that arise in discussion groups. As the examples used show, they come from many different aspects of our lives, from the banality of the kitchen to the aesthetics of a sunset. But when discussion starts arbitrarily from the questions people choose to ask, rather than the orderly pages of a textbook, what kind of a picture of science emerges in the long run? Is it a patchwork of random gobbets of information? Is it thick with context but thin on theory? Do patterns emerge? In this chapter we look at some of the transcending themes that emerge over the longer term, when hundreds of questions have been posed and countless contexts explored.

It turns out that interesting patterns have indeed emerged in open-ended discussions over the course of time and these provide interesting insights into the way people with little scientific background respond to scientific explanation. As you might expect, themes that are familiar in traditional school subjects recur regularly — the brain, digestion, electric current, the periodic table, for example. But other issues also emerge that do not lie within any one scientific subject area, instead they transcend several and, as such, may not be addressed directly in any one of them. In the following pages a number of these are described, based on issues that have recurred with different groups in a variety of contexts. Taken together they seem to occupy a kind of intermediate zone between the untutored thoughts about the natural world that occur to people as they go about their business and the codified knowledge of formal scientific disciplines.

Themes

Size and scale

One of the most important recurring issues, whether in relation to the human body, the components of matter or a journey to Mars, is the relative size of things. So many important things talked about today are microscopic, but for most people it is difficult to grasp just how small things are relative to one another — how small is small? Does a virus fit inside a cell; is a molecule bigger than an atom; is DNA bigger or smaller than a microchip? Then there's the opposite extreme — how big are stars; how many are there in a galaxy; how large is the universe itself?

It can be comforting to discover that none of us really has a perfect, intuitive sense of these extremes of scale. It is more or less impossible to visualise something a billion, billion times smaller than something else; our brains have not evolved to achieve such mathematical feats. They are better suited to comparing the sizes of objects our eyes are capable of seeing and our limbs capable of grasping. But ever since early philosophers tried estimating the size of the Earth or the distance to the Moon we have needed methods to compare sizes mathematically on a wider scale.

One way scientists capture this huge range of sizes is by talking in terms not of exact sizes but of "orders of magnitude" — things are "of the order of" centimetres, or of metres or kilometres, for example. So a thing that is twenty centimetres long, like a notebook, is of the order of tens of centimetres; a thing four metres long, like a car, is of the order of metres. In this way we can say that the size of the everyday objects we normally handle ranges over about three orders of magnitude — ones, tens or hundreds of centimetres. It is, of course, no accident that things we handle are of these orders of magnitude because our bodies, our very fingers and thumbs, are on this scale too.

There are, of course, things we encounter which lie beyond this range. A mountain might be 3,000 metres high, a human hair one millimetre in width. We are clearly able to comprehend, if not to quantify, magnitudes over a wider range. To the naked eye we are able, given good eyesight, to distinguish minute things about one tenth of a millimetre across and also to observe grand panoramas covering distances of many tens of kilometres. Thus the orders of magnitude we can readily comprehend, even though beyond our handling, extend downwards to millimetres and tenths of millimetres and upwards to tens, hundreds, thousands of metres (kilometres) and tens of kilometres — effectively a further six orders of magnitude. As a result, we get accustomed to comprehending, in our everyday lives, roughly nine orders of magnitude.

However, through scientific development we have become capable of appreciating orders of magnitude much greater and much smaller than the nine we ordinarily deal with. This possibility has come about largely as a result of the invention of instruments capable of extending the range of human vision. The microscope has opened up the world of minutiae and the telescope, the farther reaches of the universe. These two technological advances, originating in the late fifteenth and early sixteenth centuries, had revolutionary effects on the way we saw ourselves as human beings — as part of a much more extreme world than previously imagined. We learned to see ourselves as coexisting on Earth with miniscule life forms and to see our Earth itself as located quite arbitrarily in a vast universe of other worlds.

Most of us do not find it at all easy to imagine the difference between, say, a human blood cell, a hundredth of a millimetre across, and a virus, a mere thousandth of a millimetre across — they both sound absolutely minuscule! But scientists have to cope with such differences routinely — after all, one is ten times bigger than the other. They simply get on with their tasks of measuring minute and gigantic things methodically, recording them, talking about them and drawing human-scale pictures to help in visualising them. For the vast majority of us, however, differences of size on the very small and very large scale cannot be readily grasped. Yet to get an insight into what things are made of, to imagine the inner structure of substances or to visualise the constant activity within living matter, we need to try to do this. Fortunately, the concept of "orders of magnitude" is there to help and in recent years the internet has provided us with some extremely helpful devices to take us through them. Take, for example, Google Earth or any of the map tools that enable you to zoom in or out. With a simple slider you can control whether you look at a country, town, street or house — and scale each to the size of your screen. Zooming from your house to the whole of the Earth takes you through six orders of magnitude. Now there are internet tools that enable you to extend this to all scales in the universe. One such, called *The Scale of the Universe* (http://htwins.net/scale2) is a sophisticated zoom device that takes you all the way from the particles that make up atoms to the galaxies that structure our universe. Not only does it provide images of atoms, molecules, cells and stars to help you compare sizes visually, but there's also a scale, as with a map, that helps you see where you are in absolute terms.

So, equipped with the concept of "orders of magnitude", let us now look at the next regularly recurring theme: the structure of matter at different levels. In the next section we move down from the scale of everyday objects to the organs and cells, molecules and atoms, nuclei and quarks of which all substances are composed. Hold on to your seat belts!

Structure at every level

Much to the surprise of their earliest users, microscopes and telescopes reveal that matter is highly structured at every level, from atoms to galaxies. The very existence of microorganisms came as a surprise to the first users of microscopes as did the moons of Jupiter, revealed by Galileo's newly invented telescope. It turned out that the materials we are familiar with are not simple aggregates of zillions of elemental atoms of themselves — wood atoms, granite atoms, liver atoms. Instead, between an object and its atoms lie many levels of organisation.

The way matter is organised beneath the visible level proves equally fascinating for people today. A helpful way of exploring this is by way of analogy: we might, for example, imagine the levels of structure within a city. You might start by saying it was composed of thousands of basic units called buildings linked together by long thin roads, railways and paths. They are clustered together at a higher level into neighbourhoods, and these in turn into towns and cities. At lower levels, buildings themselves have a sub-structure of rooms and corridors and, within these, items of furniture and within these drawers, legs and so on.

With this image of limitless levels of structure, ascending and descending, the notion of defining a *fundamental* entity becomes a matter of choice — is it to be the building, the room, the furniture? In a similar way the notion of an atom, conceived by the ancient Greek philosopher Democritus as an irreducible, fundamental entity, though prescient, turned out to be illusory. What we today call the atom, despite its origin as an ancient Greek word meaning "uncuttable", is simply one of the many levels at which matter is organised. The exploration of levels below it continues today.

A first step in understanding the fundamentals of matter must be to identify the various levels of structure and to get a grip on how big and small different things are beneath the visible level. Where we are unable to see things with the naked eye, it is all too easy to condense their various sizes into a single "microscopic" scale — that is to say, one that is only visible through a microscope. But in reality these tiny sizes differ from one another as much as the size of your waist differs from that of the Atlantic Ocean.

What the microscope has revealed about the very small, the telescope has done at the other end of the scale. Since its invention 400 years ago it has explored the many levels of structure beyond the Earth, Sun and planets with which we are familiar. Today, researchers are still discovering more about the structure of the cosmos with the additional help of telescopes that respond to non-visible signals, such as radio and infrared waves. The broad concept of a system of orbiting planets around a central star has been understood for a number of centuries, based on our local system with the Sun at the centre. The recognition that the Sun is in fact just a star, very much closer to our Earth than all the others, followed. During the twentieth century the number and nature of stars gradually became clearer as the power of telescopes grew. We now understand them to be distributed over a vast range of distances from us, stretching to the farthest reaches of the universe. They are of different sizes and at different stages of their life cycle, but all are stars. They are not, however, evenly or randomly distributed; a higher level of structure appears to exist. Later in this chapter we look in greater detail at the nature of structure at both the microscopic and cosmological levels.

Living and not living

By thinking about the levels of structure in materials we are led to consider the distinction between living and non-living substances. The leather of a handbag is from the hide of an animal, composed of once-living cells; the metal of its buckle, a crystalline array of atoms grouped in grains; the polyester of its lining, long fibres of polymer molecules. Some materials come from former plants or animals, others from ores mined from the Earth, others from the vats of the polymer industries. Whether substances are of living or non-living origin is one of the sharpest distinctions people draw. Intuitively, living material is felt to be of a different nature from the mined or artificially created.

The strength of this intuition makes discussion about the chemical basis of living material particularly interesting. A stark example of this arose in the discussion about atoms and elements outlined on page 70 in which a participant said: "What do we mean by a compound thing like, say, a daisy — it has petals, flowers, stalk, etc. The daisy eventually dies but don't the atoms live on?" As this suggests, when forced to think about it, we acknowledge at some level that the most fundamental components of matter — atoms — transcend the living/non-living boundary. The atoms that make up a daisy continue to exist when the daisy is no longer alive, they are present in the humus that remains and may end up in another plant that springs up from it later. So essentialist beliefs, held throughout history (and to some extent today), that substances are made of essences of themselves have been replaced in modern times by an understanding that substances are composed of atoms that are identically the same whether they occur in living or non-living matter. As forms change at higher levels of organisation, the atoms themselves remain unaltered.

Central though this understanding is to modern science, it is not so easy to accept in all its ramifications. After visiting the Coalbrookdale Museum of Iron in Shropshire, England, where he witnessed the eighteenth-century processes for smelting and purifying iron, a member of one discussion group said, "It's extraordinary to think the iron we know as a strong metal is the same as the red material in our blood — they seem to be utterly different kinds of substance, with different characters." The realisation that the iron in our blood is the same stuff as the iron in railway lines and that both derive from the ores in the rocks of our Earth is typical of the many "epiphany moments" people experience in getting to grips with scientific ideas. The idea is confronted time and again as we see that the acid in our stomach (hydrochloric) is the same as we use in household cleaning; the sodium and potassium that drive our nervous systems are the same as in the salts we eat. In short, our scientific understanding puts paid to the intuitive idea that some special "life force" or extra ingredient is present in living organisms — a major challenge to our habits of thought, and indeed a spur to the much more difficult question: what is it that accounts for the phenomenon of life?

One particular challenge to the definition of "living" is the case of viruses. A discussion was once triggered by a fear that travelling by aeroplane exposes you to risk of infection. Exploration of people's experiences of infection and their understanding of its causes led to the simple question "What is the difference between viruses and bacteria?" As doctors keep reminding us, antibiotics only work for bacterial infections, not for viral ones. Bacteria are known to be living organisms comprising a single cell that contains all the machinery necessary for its survival and reproduction. Viruses are much simpler entities, assemblies of giant molecules — proteins and nucleic acid — that reproduce by attaching themselves to cells and hijacking their machinery for their own reproduction. For this reason, though bacteria are acknowledged to be alive, for viruses the case is not so clear; in the words of one expert[i] they "are thought of today as being in a gray area between living and non-living". So viruses and bacteria challenge what we mean by living and non-living. They also remind us of the question of scale: though we think of them both as microscopic, viruses are generally even smaller than the bacteria upon which they rely for reproduction — a bacterium is roughly a thousandth of a millimetre and a virus is typically around ten times smaller. The question of size introduces a further important issue that crops up in many discussions about the natural world: the distinction between the visible and invisible.

Visible and invisible

In the most straightforward sense, the word "visible" refers to things we can see with the human eye. However, the invention of the microscope profoundly affected this simple definition. The first microscopists began to see unexpected detail in living material — "male animalcules" (spermatozoa) thrashing around and blood cells, glandular secretions and muscle fibres, for example. As one historian of science[ii] has put it: "Microscopic analysis of the tissues of animals aided the conception of the living body as a mechanism." Our sense of what is visible depends intimately on the technology available.

Optical microscopes (i.e. ones that use ordinary light to illuminate the sample) make the basic structure of small organisms, such as bacteria, yeast and fungi, visible (see image, "A microscope image of a bacterium called *E. coli*" on the following page).

For smaller entities, such as viruses, a more powerful microscope is needed. The transmission electron microscope, invented in the latter half of the twentieth century, uses a stream of electrons in place of visible light, to reveal much finer detail (see image, "Yellow fever viruses, using a transmission electron microscope"on the following page).

For even smaller things, such as the giant molecules — DNA and proteins — illumination with radiation of much shorter wavelength than light is needed and X-rays provide a means for doing this. With the aid of a

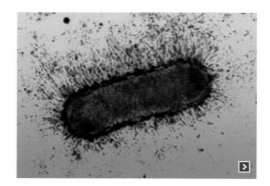

A microscope image of a bacterium called *E. coli*[iii]

Reproduced by kind permission of David Low, Edward N. Robinson, Jr., Zell A. McGee and Stanley Falkow, University of California, Santa Barbara

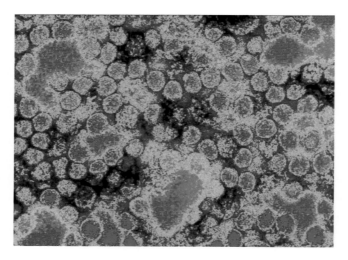

Yellow fever viruses (coloured green), using a transmission electron microscope[iv]

Reproduced by kind permission of the Centers for Disease Control and Prevention, Atlanta, GA, USA

computer, a model of the molecule can be constructed from the pattern X-rays make when they pass through a crystalline form of a substance. The image "Molecular model of polio virus receptor" shows a model of the large protein molecule that a polio virus attaches to in the body. Each sphere represents an atom in the molecule.

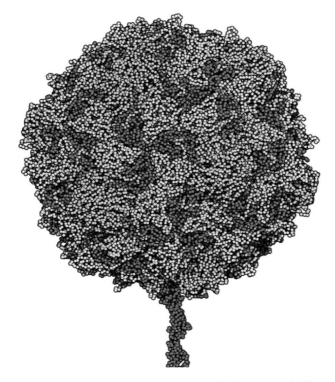

Molecular model of polio virus receptor (schematic model)[v]

In recent times it has even proved possible to create images below the level of the molecule. Direct images of atoms have been made using a new technique called "atomic force microscopy". The next image shows carbon atoms arranged hexagonally in a sample of graphite.

So, direct images of smaller and smaller entities have been made possible as microscope technology has developed.

There are, of course, other kinds of invisible things that play a major role in science, which are not simply small, but immaterial; such things as energy, forces and fields, for example. The ability to physically see things plays an important part in the effort to understand a scientific phenomenon. The use of microscopes greatly helps by enabling us to see images of viruses, molecules and even atoms. Abstract concepts, so fundamental to science, are equally hard to grasp so long as they can't be visualised. A magnet clinging to a fridge door may be highly visible, but the influence that makes this happen — the

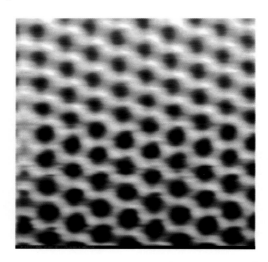

Graphite with carbon atoms arranged hexagonally[vi]

Note: The picture is two millionths of a millimetre square

Reproduced by kind permission of Experimental Physics VI, University of Augsburg, Germany

magnetic field — is not; similarly for the electromagnetic wave that delivers our mobile (or cell) phone messages. But abstract concepts such as fields and waves come to life when made "visible" through a diagram or physical model that helps make them concrete.

Taken together, these examples of structures and influences that we are unable to directly see, demonstrate an important aspect of learning about science. Our intuitive understanding of invisible structures and mechanisms is based largely on that which we can see or touch. This helps us imagine the invisible and enables us to create metaphors to aid our thinking. However, the urge to visualise can also act to limit understanding. Many aspects of modern science depend on willing suspension of the visual. The effects of relativity theory on our notions of distance and time, and of quantum mechanical explanation on the behaviour of matter, take all of us, scientists included, beyond the realm of the purely visual. Ultimately, mathematical explanation prevails where metaphor fails.

Having identified a number of transcending themes emerging from discussion of many different aspects of science — the relative size of things, levels of structure, living and non-living and imagining the invisible — we now turn to consider one of these in greater detail: the way in which matter is structured at different levels.

Structural Components

As we have seen, structure exists at every level of matter. In this section we draw on a wide range of discussions that have taken place about the human body and everyday materials to explore some of these structures. They range from the biological cell to the fundamental particle.

Cells

Amazingly, almost all living material is made of cells. Some, such as red blood cells and nerve cells are familiar, featuring regularly in everyday language. A few, such as stem cells or egg cells, even make it to the news from time to time. But the vast majority of the countless different types that make up our bodies remain barely known at all to the layperson. They vary tremendously in size and shape and perform an extraordinary diversity of functions, but have enough features in common to be called the same thing — cells. Typically they range in size from a tenth to a hundredth of a millimetre (10^{-4} to 10^{-5} metres — around ten to the width of a human hair). This means that apart from a few exceptionally large ones that can be seen with the naked eye, they require a medium-power microscope to magnify them. In fact it was only after the microscope had been invented and used in the seventeenth century that people began to realise that living matter was in fact composed of a mass of cells.

So, if cells vary so much, from the humble skin cell to the exotic light-sensitive retinal cell, what is it they have in common? The key feature, which accounts for the name itself, is that they all have walls designed to keep the inside and the outside apart — much like their prison namesakes. Unlike these, however, biological cells are flexible and porous, enabling things to move in and out as required. Inside the wall is a fluid (called cytoplasm), within which smaller structures are dispersed (called organelles) that carry out the various functions of the cell. The nucleus is one of these, storing genes that contain the information needed for the cell to function and to reproduce itself. The mitochondrion is another, making energy available for

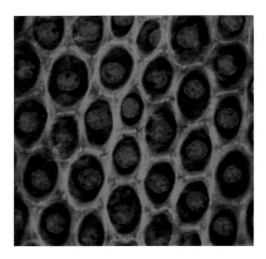

Cells behind the retina of a bovine eye[vii]

Reproduced by kind permission of Robert Fariss PhD, National Eye Institute,
National Institutes of Health, Bethesda, MD, USA

the continuous activity undertaken in the cell. Overall, a cell is a kind of active community, with its component parts working in an interdependent way, enabling it to survive in a constantly changing environment, to function effectively and to reproduce itself.

So, if a cell, at a scale of hundredths of a millimetre, is an active living community of many smaller entities, what are these entities made of? What is the material of the walls, the cytoplasm and the organelles within a living cell? What size and shape are they? To answer this, we move down a level to consider structure below the microscopic.

Molecules

The cell and its components are made of a huge range of even smaller structures that also share a common name: molecules. Like cells, these also vary enormously in size and shape, ranging from giants of a tenth of millimetre to pygmies of about a hundredth of a thousandth of a millimetre — as different in size as a house and a city. Yes, truly, molecules are as varied in size as this, so it is hard to imagine them as being of one kind — big ones contain all the information in your genes, little ones make up the air you breathe. As a very rough guide to the scale of things, molecules are typically about 1,000 times smaller than cells.

Some models of molecules of various sizes are:

A water molecule (three atoms)[viii]

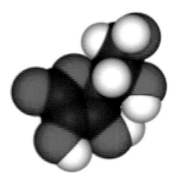

A vitamin C molecule (twenty atoms)[ix]

A stretch of a polyethylene molecule (thousands of atoms)[x]

If molecules vary so much, what is it that they have in common? What defines them? The practical answer is that they are the units of matter that don't easily get broken up in everyday circumstances. Molecules endure, rather like the bricks of a house; they are capable of being broken up, but mostly they remain as they are. In air, for example, if you could get down to the molecular scale, you would "see" endless, almost empty, space interspersed with occasional molecules of nitrogen, oxygen and carbon dioxide buzzing around at high speeds. These molecules are relatively small as molecules go, being about a tenth of a millionth of a millimetre in diameter. In other words, about ten million would fit across the width of a human hair. At the other end of the scale of molecules is DNA, which, if you could get down inside the nucleus of a cell, you would "see" as a long thin thread wrapped up like a ball of wool some tenth of a thousandth of a millimetre across.

In living matter there are countless different kinds of molecule. A protein such as haemoglobin is one; a lipid such as cholesterol is another; a hormone, such as oestrogen is yet another. In fact, most of the different substances we read about on food labels or medicines are examples of different kinds of molecules. Most protein molecules are globular in shape with a size of around a hundredth of a thousandth of a millimetre (10^{-8} metres) — in other words, tens of thousands would fit across a human hair. Lipids, such as cholesterol, are much smaller — ten to a hundred times so — as are many hormones such as oestrogen, and drugs such as aspirin. So molecules associated with living matter vary from the simplest, such as the oxygen we breathe, containing just two atoms, through the hormones, lipids and other medium-sized molecules in our bodies, containing dozens or hundreds, up to the proteins with their thousands and DNA with their hundreds of millions of atoms.

It is not only biological material that is organised at many levels but non-living matter too. In place of cells, a piece of metal or sand viewed through a powerful microscope reveals structures known as grains. These are zones with discernible boundaries, packed tightly against one another. Typically, they measure between a tenth and a thousandth of a millimetre across. Within each grain, molecules are arranged, often in an orderly arrangement with each molecule orientated exactly the same way as its neighbour. In contrast to the cellular organisation of living matter, molecules in such non-living substances do not move freely past one another and are not organised in flexible structures. The orderly arrangement is typical of a crystalline kind of structure, and minerals and metals are generally organised in this way. In common with living matter, however, the basic units of which grains are made are also molecules. Just as for living matter, these are the units in which matter is normally organised. In everyday life we tend to associate the word crystal with the large geographical specimens we see in museums and gift shops. These have remarkable geometric shapes, with perfect flat planes set at precise angles to one another. These are the extreme cases where the geometric arrangement is repeated so perfectly, with such fidelity over millions and millions of molecules or atoms, that the internal symmetry at the microscopic level is reflected on the outer surface of a chunk of crystal. In the less spectacular case of ordinary metals and minerals the molecules or atoms are also arranged geometrically, but the perfect arrangement only extends over a much smaller expanse — the extent of each grain. The boundaries where one grain meets another are not arranged in an orderly fashion, however — the grains are orientated irregularly in relation to one another. As a result, pieces of metal or rock don't usually look like crystals to the human eye.

There are many other distinct ways in which materials are structured at the microscopic level — as amorphous powders, liquids, gels, glasses and polymers, for example. The nature of these is the basis of the discipline of materials science, a fascinating area of enquiry but not one we will pursue further here.

Atoms

Moving down to the next lower level of organisation, molecules are themselves composed of even smaller units called atoms, each bonded to its neighbours.

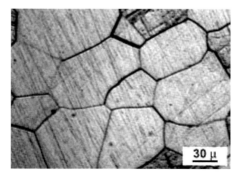

Grains and their boundaries in a metal (property of Edward Pleshakov)[xi]

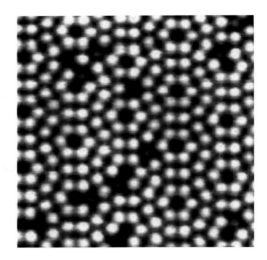

Silicon atoms, using a dynamic force microscope, by Oscar Custance. Published in *Science*, 332, 413, (2008)[xii]

Atoms themselves are pretty much of a size: they do not vary as much as molecules or cells. The smallest atom, helium, is less than a tenth of a millionth of a millimetre; the largest is only eight times bigger.

Molecules do not break up easily — this fact alone tells us that the atoms making up a molecule must be strongly bonded to each other. On the other hand, we know that molecules can be broken up if sufficient energy is made available for a chemical reaction to occur. Reactions occur all around us all day long: the enzymes in our stomachs break up the molecules of our food; the steel of our cars rusts slowly over time; the leaves of trees and plants use carbon dioxide from the air to synthesise the molecules they need. But as molecules break up and re-form in different combinations, the atoms of which they are composed remain unaltered — they simply transfer from one molecule to another as the reaction takes place. Oxygen atoms from the air are incorporated in the iron oxide of rust; carbon and hydrogen from the carbohydrates you eat make up the molecules of your body. In normal circumstances, atoms themselves do not break up.

Below the atom

But what of atoms themselves? Are they indivisible, as the ancient Greeks imagined them to be, or are they composed of particles of an even more fundamental kind? This question was answered in a way that all could see when the atomic bomb made its unhappy entry upon the world stage — the atom could indeed be split.

Here on Earth, atoms do not ordinarily break into bits — indeed, it would be a lifeless, radioactive planet it they did. But they can be made to, and when they do, unprecedented concentrations of energy are released. The breaking up of an atom is called "atomic fission" and it occurs on Earth in nuclear power plants and nuclear explosions (and in a few naturally occurring radioactive ores). So atoms are indeed made up of bits; however, as they mostly remain intact indefinitely it is clear that these bits must stick together very tightly under normal circumstances. After all, atoms come unscathed through chemical reactions of all kinds. Sulphuric acid may leach away at a piece of steel and dissolve it, but the atoms of sulphur, oxygen and iron remain identically the same before, during and after the reaction. A few kinds of atom, however, do actually break up spontaneously — uranium is a naturally occurring example. Others can be made to break up by shooting fast-moving particles at them. This is what happens in a nuclear power plant or bomb.

So if atoms can be broken up into parts, further questions are raised: what are these more fundamental particles; and what size are they?

The quest for a model of the atom was an exciting part of scientific discovery around the turn of the twentieth century. Many different scientists from various countries conducted ingenious experiments that together revealed that atoms must be made of a mixture of negatively and positively charged smaller parts — which came to be called, unsurprisingly, particles. As we know from daily experience, everyday objects are not normally charged up electrically — they are neither positive nor negative on balance, so we conclude that the positive and negative particles must be exactly equal in number.

In a revolutionary experiment in Cambridge, England, in 1911 the scientist Lord Rutherford showed, to everyone's great surprise, that the inside of an atom was overwhelmingly … empty! After recovering from the shock he concluded that the particles of which it is composed must be extremely small. He showed that the positive particles must be concentrated in an unimaginably small space in the centre, which he dubbed the nucleus (from the Latin for the kernel of a nut). This nucleus is as small in comparison to the whole atom as a speck of dust is to the Royal Albert Hall — a mere 0.001% of the atom's volume. Although it is so small this nucleus nevertheless contains 99% of the mass of the atom. It comprises two main types of

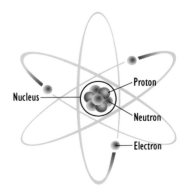

Structure of an atom (property of Fastfission)[xiii]

massive particle — positively charged ones called protons and uncharged ones of similar size and mass called neutrons. The negative particles balance the positive ones exactly to ensure overall neutrality but are, by contrast, tiny and ultra light; they are called electrons. Each carries precisely the same amount of negative charge as the positive charge on a proton, but they are located far, far away from the centre — in fact, their distance from the nucleus is about a thousand times the size of the nucleus.

Below the particles

The discovery that atoms were not, after all, the smallest unit of matter, drives us to reiterate the fundamental question — are the particles that make up atoms made of even smaller units? Indeed, this question taxes today's scientists in both theoretical and experimental ways. A widely held theory proposes that particles are indeed composite, each made up of even more fundamental entities called quarks; and the quarks themselves come in a variety of different types. From the point of view of size, quarks, though they have never been directly detected, are considered to be smaller than 10^{-15} metres — a millionth of a millionth of a millimetre. Beyond the quark, we enter the realm of hotly contested particle physics and are forced to leave it to writers of the future to describe the size of things yet to be identified.

Cosmological structure

While scientists in one branch were exploring the make-up of matter at ever smaller scales, others were looking to the heavens and discovering structure on the grand scale. Stars and their planets have always been a source of fascination being, as they are, visible to the naked eye. Stars were seen to remain in fixed positions relative to one another, even though the heavens as a whole appear to move as the Earth turns. Planets were observed to move, however, and during the Renaissance period were understood as orbiting the Sun. The Sun itself came to be understood as simply another star, though very much closer to Earth than all the others. The Milky Way appears to the casual observer as a whitish blur in the night sky but the telescope eventually showed it to consist of a very large number of stars clustered together. It is now believed to contain approximately 100 to 400 billion individual stars. As other groups of stars like this were gradually discovered the name "galaxy" was given to them, based on the ancient Greek word for milk.

More recently it has been found that the galaxies themselves are not fixed in position but rotate about their own centre and have distinct shapes. Some are disc-like with stars arranged along the curved arms of a spiral (rather like a Catherine wheel); others are ellipsoidal (an ellipse or oval shape in three dimensions); others are more irregularly shaped. Difficult though it is to imagine, these galaxies not only contain enormous numbers of stars, but the galaxies themselves are also extremely numerous — there are thought to be more than 170 billion in the observable universe.

With our eye on the existence of structure at every level, it comes as no surprise to learn that even galaxies are not evenly spread around the universe but are instead organised into higher-level structures known as "galaxy clusters" and these in turn are grouped in to "superclusters". Research today is showing that at the highest level of organisation currently understood these superclusters appear to be arranged throughout the universe in threads known as "filaments" or spread out in "sheets". Thus the material in the universe, far from being amorphously or randomly organised, appears to be highly ordered at multiple levels. The view through the telescope has a remarkable symmetry with that through the microscope — matter, it turns out, does not appear to be aggregated in arbitrary, amorphous ways but to be structured at every level.

Systems of Classification

Getting into science is something of a journey. New worlds begin to open up: the microscopic and the vast, the living and the inanimate, the tangible and the abstract. Questions that at first seem simple — what is a hormone? Why is the Earth magnetic? — spark off enquiries that reveal ever more complexity. It's rather like approaching what looks like the peak of a mountain only to find it is just a minor outcrop on the way. Science, in its entirety, can seem overwhelmingly huge and impossibly complicated; there is so much knowledge to absorb and so many conceptual barriers to break through. This perception can be daunting but it shouldn't be off-putting; indeed, many scientists feel overwhelmed by the complexity outside their specialist field. However, through their training scientists acquire ways of organising this vast body of knowledge, which can also be a great help to the layperson. These are the various systems of classification and ordering that help reduce the overpowering sense of infinite detail. Some examples of classification systems that have proved helpful in discussion groups are outlined in the following sections.

The periodic table of the elements

One of the greatest simplifications in all science was the creation of the periodic table of the elements by Dmitri Mendeleev in 1869. He noticed that if the elements known at the time were arranged in order of weight, their chemical properties tended to recur in sets of eight. From this he was able to construct a table that showed the elements falling into clear families, such as the halogens and alkalis, and lined up in columns. The power of the table was demonstrated when gaps in it predicted the existence of as yet unknown elements which were subsequently discovered. The case study on pages 107–108 shows the course of a discussion about the periodic table.

Appreciation of the periodic table of the elements immediately provokes further questions: how do you *explain* the regular patterns in the periodic table? Why are all the elements in column 18 inert? Why do those in column 17 tend to react with those in column 1? In discussion groups today, as in scientific circles at the time, these led into the fascinating story of the internal structure of the atom. The example on page 109 shows how the systematisation introduced by the periodic table and the model of the atom enabled many kinds of chemical reaction to be explained.

A System of Classification

The periodic table

A person in one discussion group talked about her mother's enthusiasm for a TV documentary on the periodic table. The programme had explained that Dmitri Mendeleev, the person who had created it, was not especially clever and he had not come from a well-to-do background, but his mother had been determined to get him a good education. In 1869, when he was in his mid-thirties, he suddenly saw a way of explaining the relationship between the various elements. His periodic table of the elements was the result.

His work built on the laborious work of many natural philosophers and scientists (and alchemists) before him who had been gradually purifying substances for hundreds, even thousands, of years. They were trying to find out what was "elemental" about substances — i.e. the point at which they could not be broken down further into components. Someone in the group recalled that the ancient Greeks believed that matter was composed of combinations of four elements: earth, air, fire and water. Another pointed out that this view was not based on experiments; in fact, it was not until the Enlightenment period that careful measurements began to be made in experiments that isolated, purified and weighed these elemental substances.

It was found that equal amounts (e.g. a cupful) of different elements didn't weigh the same. This meant that the atoms that made up the substance must also have different weights. It was found that carbon atoms weighed six times more than hydrogen atoms and oxygen eight times more. Mendeleev, and one or two others before him, had the idea of laying out the elements in a sequence, from the lightest to the heaviest.

At the same time, through countless detailed experiments it also became clear that some elements had things in common. For example, some were metallic, some not; some were very reactive, others were inert. It needed a moment of creative brilliance to see how the elements could be arranged in a way that explained the common characteristics. Mendeleev did this.

At this point discussion turned to the fortunes of history in science. It seemed that Mendeleev, whatever his abilities and imagination, happened to be on the scene at just the right moment. It had taken hundreds of years of painstaking experiments by many others to accumulate the information upon which his work rested.

(Continued)

(Continued)

Mendeleev's insight was not only to lay out all the known elements in order of increasing weight but also to make breaks at certain points, placing each sequence in a row underneath the previous one. He created a two-dimensional table out of a long row. In this way he was able to place elements with similar properties underneath each other, *in the same column*.

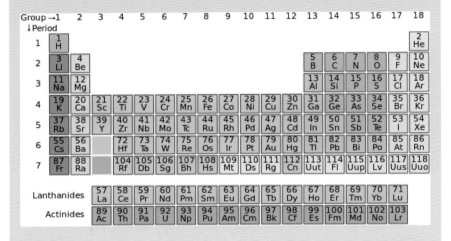

Thus, for example, the column, on the right-hand side (18) contained all the elements that were inert — they did not react with anything (helium [He], neon [Ne], argon [Ar], krypton [Kr], etc.) On the left-hand side was another column (1) that contained very reactive alkali elements — lithium [Li], sodium [Na], potassium [K], etc.)

Chemical groups

Systems of classification also help at levels above the elements. Most of the substances we come into contact with are compounds whose molecules contain many atoms of different elements. Familiar examples are carbon dioxide (CO_2) and water (H_2O). However, more complex molecules are not merely

An Example of a Powerful Model

Inside the atom

Around the turn of the twentieth century, Lord Rutherford showed that the atom contained a precisely equal balance of negatively and positively charged components. It turned out that these were not arranged in a kind of solid "pudding" of particles but as a tiny core or nucleus in the middle, which contained almost all the matter of the atom and was surrounded almost entirely by empty space. The nucleus was made up of a number of heavy positively charged particles, which were named protons. In the empty space around this a number of very, very light particles called electrons were whizzing. These were negatively charged. They were described as forming a kind of cloud or blur (like propeller blades) or as tracing out a kind of shell as they move around the nucleus.

Mendeleev's pattern was explained by inventing a model of the atom consisting of shells within shells, like Russian dolls. Each shell contained a number of electrons but there was a maximum number allowed before a shell became full up. In the inert atom helium (He) the shell is complete, with two electrons. But the next element, lithium (Li) (with three electrons), is said to have two electrons that complete its inner shell, plus one extra electron extra that occupies an outer shell. The next element, beryllium (Be) (four electrons), has two electrons in its inner and two in its outer shell — and so on, until you get to neon (Ne) with ten electrons. This is inert. So the outer shell of neon is complete with eight electrons (with two remaining in the inner shell).

The elements with complete outer shells were inert — they didn't react with anything. But the ones next to them in the rows of the table reacted strongly. The model explained this by suggesting that an element with one electron *extra* (like Li) will react with an element that has one electron *short* in its outer shell (like fluorine [F], which has 7). This explains why lithium fluoride occurs; similarly for sodium chloride.

So the periodic table was built up putting each element in order of the number of protons (and electrons) it contains. The rows were arranged to produce groups of elements with common characteristics in the columns.

made up of increasing numbers of atoms of different elements; they tend to be assemblies of specific groups of atoms. It is these chemical groups that form the important building blocks for the complex substances we come across.

Simpler substances such as methane or ammonia are the basis of chemical groups from which more complex molecules are built. Known by their adjectival form — methyl or amine, for example — such groups often give rise to the lengthy names by which drugs and household products are named. My jar of handwash tells me that I use butylphenyl methylpropional every time I wash my hands!

As an example, an amino acid is a compound built up from two chemical groups: an amino group, comprising one nitrogen and two hydrogen atoms (NH_2), and a carboxylic acid group consisting of one carbon, two oxygen and one hydrogen atom (COOH). These are shown in the model, "An amino acid".

Interestingly, amino acids themselves form a particular class of compounds. Each member of the class has the same basic structure but differs in just one part, labelled R in the diagram. A class of even more complex molecules, the proteins, uses amino acids as building blocks. Proteins are huge molecules made by linking together dozens or hundreds of amino acids in a long chain.

We have thus seen two illustrations of the way in which understanding of the huge variety of chemical forms can be simplified through systems of classification. The multitude of different elements can be understood more systematically through the periodic table and the immense variety of molecules made more comprehensible by the system of chemical groups of which they are composed. Many other kinds of classification system help us to simplify the immense diversity of substances, such as classes of metals and minerals, categories of polymers and proteins, and types of acid and alkali.

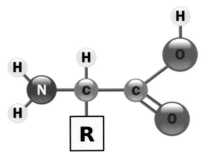

An amino acid[xiv]

Biological systems

Systematic thinking is not restricted to the physical sciences, of course — it is equally important in the life sciences. This often becomes apparent in discussions that typically begin with issues of family health and lead on to explorations of the human body. A discussion was once triggered by the back pain of one participant, caused by damage to her central nervous system. She asked, "Is the electricity in nerve cells the same as in a battery — how do nerves actually work?" This led to discussion about the bundles of fibres that we call a nerve and the nature of the specialised nerve cells of which the fibres are part. The discussion opened up a series of questions giving insight into one of the most important systems in biology, as the example below shows.

The nervous system is but one of many biological systems whose workings have been pieced together gradually over the centuries as links were made between different investigations and observations. The idea that separate organs and processes can be linked in one coherent system is helpful in learning about science as well as in the practice of medicine. A discussion in

The Nervous System

An example of a biological system

The chronic back pain suffered by one participant in a discussion group led her to ask how nerves actually work. Most people in the group understood that the nervous system is essentially electrical in nature but were baffled about how this related to our common experience of electricity through batteries or the mains. "Is it the same kind of electricity?" one asked.

The system as a whole has several linked parts, running from the nerves that reach right to your fingertips and every other part of the body, through the spinal column to the brain. This constitutes the central nervous system. Some nerves pick up sensations, from eyes, skin, stomach and so on, while others deliver signals to the parts that move — the muscles, in particular. What they all have in common is a highly specialised form of cell in which one part is extended enormously in one direction (called an axon), forming in effect a conduit for the nerve signal to travel along — in this sense it's like a wire.

(Continued)

(*Continued*)

But there the analogy stops, because unlike the flow of electricity in metal wires, charged particles do not travel down the axon from one end to the other. Instead, atoms that are electrically charged (known as ions) are pumped in and out of the wall of the nerve cell along the length of the axon in a kind of Mexican wave. So an electrical pulse occurs at the point where charged ions pass out of the cell. The pulse then moves down the axon and the ions return back through the cell wall to reset the axon at that point to its original state. The pulse moves down the axon rapidly, but much more slowly than the current in a copper wire.

In discussion this explanation unleashed a spate of further questions: What happens when the nerve cell ends — how is the message communicated to the next one? Is the connection between cells just one-way? This led rapidly to a deeper exploration of the synapse — the gap between one cell — and the next and the way in which the signal passes across, not by electrical but by chemical means. This in turn inspired the group to find out more and to see what is being researched today. A visit to a cell biology lab was arranged, which not only showed how synapses work in general but how drugs might be designed for use at the synapse to help people with clinical depression. This condition appears to be linked to the functioning of serotonin, a vital chemical transmitter at the synapse.

one group began with the historical role of the microscope and led on to the realisation that the circulatory system of the blood had only been established once the very fine capillaries had been discovered that enable nutrients and waste products to be exchanged between blood and tissues. The hormonal or endocrine system is another concept that links things together and simplifies understanding. The connection between hormones that are produced in one part of the body, often a gland, and the effects they give rise to elsewhere is important in helping make sense of apparently unrelated things. One memorable discussion focused on the pituitary gland in the base of the brain and enabled participants to see the link between nerve impulses in the brain, the release of hormones into the blood stream, and the consequent bodily effects. There are many other systems in the body, e.g. to regulate body temperature, control blood sugar levels, fight off infection and transfer energy from food to muscles. Insight into how they operate as a whole helps bring order to the mass of detail when learning about the workings of the body.

This foray into organised structures of knowledge, as diverse as the periodic table of the elements and the human nervous system, illustrates an important point for people wishing to engage more with scientific ideas despite having little background knowledge to begin with. When curiosity drives us to ask a specific question, it may not be that we are really looking for an equally specific answer to it. It may more usefully be the point of entry for learning about some whole system of knowledge. Equipped with insight into such systems we place ourselves in a more powerful position to explore and understand the apparently endless complexities of the natural world.

Challenging Ideas

In this chapter we are considering some of the underlying issues that arise when people start exploring scientific ideas by drawing on their experience of everyday life. We have identified a number of transcending themes that emerge — size and scale, the invisible, living and non-living — and considered some ways of dealing with complexity, through levels of structure and systems of classification. We now turn to a further set of challenges, the intellectual ones that arise when grappling with concepts that are just hard to understand (or at least to accept).

Movement

One of the most fundamental of these challenging ideas is also one of the longest standing: what it is that makes things move. Aristotle was clear about this: for him things move according to the elements they are composed off — smoke rises because it is air-like; heavy objects fall because they are earth-like. He taught that heavy objects fall faster and that the speed of an object depends on the force being applied.

Some of these theoretical ideas match quite closely our everyday experience of moving things, so it is hardly surprising that for many people today they seem true: heavy objects do usually fall faster than light ones. In recent years, though, it has been possible to actually see that however true this may appear on Earth, elsewhere it simply isn't. In outer space, far from gravitational forces, a feather really does fall as rapidly as a hammer, free from the resistance of the Earth's atmosphere. A classic video by astronaut David Scott, available on YouTube,[xv] shows this, dramatically, in the vacuum of an orbiting space station.

With the Renaissance, experimentation was gradually introduced into science where previously reasoning alone had sufficed. Galileo studied motion by rolling balls down tracks of varying steepness, and in so doing laid the foundation for Newton's profound theories about motion. Today, with video images from outer space, we are in a better position to grasp the points Newton was making, points that eluded so many of us at school! Objects just keep on moving steadily as they are if there is no force on them. In outer space people and objects just keep on travelling indefinitely until they hit an object. No force is necessary.

The idea of "force" as something that causes things to speed up or slow down or change direction was expressed clearly by a person in a discussion about Newton's theory: "You can certainly feel the effect of acceleration — when you pull away from traffic lights in car, you sink back in your seat; or when a fast train takes a curve you are pushed to one side." Another person added, "When an aeroplane suddenly descends in a pocket of low pressure air, or when a lift starts to descend, you have a sinking feeling in your stomach." But one of the difficulties in comprehending the force due to gravity was put honestly by another participant: "Gravity seems too 'negative' a thing to be called a force. It just lets things happen passively — it just makes things fall. It's like an absence of something."

Relative movement brings a further intellectual challenge — what happens when two things are moving in different ways? One discussion focused on the motion of someone walking through a moving train. If you pass slowly by in a faster train you sense them moving backwards. Some in the group pointed out this was only relative to you on your train; the other person knows they are walking forward and they see you moving ahead of them even faster. But we can tell the true situation because we have the surrounding countryside rushing past, reminding us which is moving fastest. With this tentative grasp of ordinary relativity, discussion moved on to imagining an idealised situation in which your train was moving perfectly smoothly at a steady speed in a straight line, so you were no longer aware of its motion at all; and the surrounding countryside was blanked out so you couldn't see yourself rushing past. Then who could say which was moving: you or the person on the other train? This "thought experiment" become the entry point to an exploration of Einstein's special theory of relativity, based as it is on the realisation that there is no way of saying which frame of reference is moving fastest. There is no absolute meaning to the concept of "stationary".

Action at a distance

An equally taxing concept for us today, much as it was for ancient and medieval philosophers, is how one thing can affect another without the two being in direct physical contact. So long as we restrict our thoughts to discrete mechanical systems like a bicycle or a human being, there is no problem: we extend a leg, push on the pedal, the chain moves, the wheel rotates. But how can we explain action taken in one place that has an effect somewhere else? How does a magnet on a table attract a pin ten centimetres away? How does heat from the Sun, millions of miles away, evaporate a puddle here on Earth? In other words, how do we explain action at a distance?

In the fourth century BC, Aristotle tackled the astronomical aspect of this problem by postulating a series of concentric crystalline spheres with the Earth at their centre. The Sun, planets and stars were each attached to one of these spheres, which were able to rotate, one within the other: a mechanical solution, in which movement of one physical object was assured by its attachment to another. Even 2,000 years later, Isaac Newton expressed discomfort with the idea of one thing simply acting on another without a physical link between. As he said in a letter:

> That one body may act upon another at a distance thro' a Vacuum, without the Mediation of anything else, by and through which their Action and Force may be conveyed from one to another, is to me so great an Absurdity that I believe no Man who has in philosophical Matters a competent Faculty of thinking can ever fall into it.[xvi]

It is little wonder that the same disbelief is prevalent today — how can actions be communicated over a distance without an intervening mechanism? Yet the issue is perhaps more troubling today, because action over distances is so much part of everyday modern life. We pick up a mobile phone — it is apparently physically unconnected to anything, yet it rings, it transmits our voices. We watch a Mars mission on TV — the vehicle moves, yet the instruction to do so had to pass through the vastness of the void to reach it.

So what does the job of connecting rods and levers when it comes to effects that travel over distances with no intervening mechanism? To cope with this huge conceptual problem, scientists developed the concept of "fields" — gravitational, electrical and magnetic fields, in the first instance, to explain the observed facts of attraction between things. Heavy bodies like the Earth and Moon stay in perpetual embrace; a magnet and a piece of iron are attracted to one another; positive and negative electric charges attract or repel. The field concept simply describes what is observed as a given fact, without attempting to explain it in terms of something more fundamental. The given fact is that one thing just *does* attract (or repel) another thing over a distance.

Thus, around a given body there exists a field (or fields) of influence. Around a magnet is an influence that attracts pieces of iron and attracts or repels another magnet. Around a charged-up comb is a field of influence that attracts any fine hair it may have just run through. And, more abstractly, around any object lies a very, very weak field of influence just due to the matter within it — a gravitational field; though this is so weak it is only noticeable for massive bodies such as the Earth, Moon or stars. And these areas of influence, or fields, extend indefinitely outwards in all directions from an object and happily traverse empty space.

Armed with this concept, attraction is more easily explained. If there is a field of influence surrounding say, a magnet, and it extends indefinitely, then any piece of iron sitting in this sphere of influence will feel a force exerted on it by the field; and this force may move it — either repelling it away or attracting it nearer. Again, the mystery of an underlying cause is not "explained" by this, but the model describes the process in a powerful and predictive way. Similarly for the Earth and Moon, or any two massive objects: round each object is a field of gravitational influence and each object in turn experiences a force from the field around the other object. The mystery of how the influence spreads across space is simply sidestepped, with the assertion that it does so; its effects can be observed and the "field" concept predicts very precisely the nature of the effects.

Once the concept of a field had been established and proved useful, the way it travels over space became the subject of debate. In the nineteenth century a theory was developed by the Scottish scientist James Clerk Maxwell that fields travel outwards from their source, not as a kind of bullet but as a kind of wave — a vibration that moves forward, rather as the ripples on a pond move forward, even though the water at any particular place simply wobbles up and down. It had already been established that whenever an electric charge moves it gives rise to a magnetic field in addition to the electric field that surrounds it. If an electric charge starts to vibrate up and down it causes these two fields, electric and magnetic, to vibrate in tandem with each other and the two linked fields move outwards as a wave, much as a wave machine at a swimming pool sends out continuous ripples across the pool.

Strange though this explanation may sound, it forms the basis of so much we experience in everyday life. It is the vibrating electric charge in your aerial that gives us TV and mobile phone signals; it's the vibrations in a hospital X-ray machine that create the waves that penetrate our skin and reveal our bones. These waves, known collectively as "electromagnetic", permeate space all around us and are fundamental to our existence: they make life visible, warm and comfortable for us human beings.

Interestingly, other more human forms of action at a distance crop up from time to time in discussions about science in everyday life: the actions that take place daily in our bodies. Pain is felt in the brain when a finger is cut; hunger is felt when food is lacking; leg muscles are activated when your brain senses a threat. These types of action, however, are generally not mediated by electromagnetic waves but are stimulated by the movement of physical substance between the source and the active point. Two key mechanisms mediate activity in the body: nerves and hormones. Nerves are activated by, say, pressure on the skin, or light impinging on the retina. Such stimuli cause physical movement of molecules and atoms within the nerve cells and these movements are communicated along the length of nerve cells and on to the next one until they reach the central nervous system and activate it. Hormones, on the other hand, are chemicals manufactured and stored in specific parts of the body — the glands — and are released into the bloodstream when the gland is stimulated. They move physically through the body until they reach their target site where they interact with other molecules and produce a particular effect — lactation at the breast or speeding of the heart, for example.

So actions take place over distances in a variety of ways. In simple cases, this happens by direct mechanical contact, as in a bicycle, or in biological systems by the flow of chemicals along channels. In more mystifying situations, such as light reaching our eye from distant sources or moons circling round their local planet, the action occurs through a mediating field that permeates all space.

These reflections on the minutiae of human physiology and the mysteries of outer space lead us to the final theme of this chapter — the extreme nature of so many scientific phenomena.

Extremes

As we sit round in a wine bar talking about scientific ideas or sit at home watching a documentary we are surrounded by the ordinary objects of the world we inhabit: chairs, tables, telephones, people, clothes, cars and so on. Varied though they are, all the objects we normally interact with have something basic in common: their scale and lifespan. Of course, I don't mean things are all similar in size to within a metre or last exactly the same number of years; we are talking about approximate average values — orders of magnitude. Only rarely do you come across something 1,000 times bigger than yourself in everyday life or something that has lasted 1,000 times longer than you. On the whole, the dimensions of most objects we encounter regularly are of the order of a few metres or centimetres, though occasionally we might come across a skyscraper or an ocean liner. Similarly, most things we deal with in everyday life are a few days or years or decades old, though occasionally we will witness a microsecond flash or enter a thousand-year-old cathedral. The world around us seems almost to be scaled to us as human beings — a couple of metres of height and a few score years of life. Or could it be the other way round? Have we evolved as we are in order to fit the scale of the environment we find ourselves in? Either way, the consequences for our ways of understanding the world are the same: we seem to feel most comfortable with concepts matching the scale of our environment. When we pose our scientific questions we think mostly of things on a regular scale: brains and nerves, cars and fridges, clouds and rivers. Similarly, we tend to talk most readily about things in the world that move at reasonable speeds, are reasonably nearby and can be readily touched and seen or heard by us. In other words, the sense we gain of the natural world is conditioned by our human bodies — our dimensions, our sense organs, our lifespan. The point was captured nicely by a joke told in a discussion group:

"How old is the dinosaur in this museum?"
"Twenty-four million and three years."
"Why the three?"
"Well, I came here three years ago and it was twenty-four million then."

Yet as soon as we make this point, it is easy to see how limiting, or even misleading, this perceptual framework can be. We know that the natural world presents timescales and dimensions over a vastly greater range than we ordinarily encounter as human beings; and phenomena that can't be sensed by sight or touch are all around us; and things happen faster than we can record. The world is much more extreme than we at first think. However, for science these extremes are all in scope: the natural world in its entirety — from the boson to the black hole.

With this in mind it is little surprise that discussions that start with back pain or earthquakes, diets or cell phones, quickly lead to processes on scales that simply astound the mind. In coming to terms with fundamental concepts we are often confronted with extraordinary figures or unbelievable processes that leave us speechless. "What do you mean: how can it be so miniscule, so far away, so rapid, so invisible?" In trying to comprehend such extreme values we need to go easy, allow time for the magnitudes to sink in and comparisons to be made with things we know and understand. As well as trying to understand tiny numbers, like a millionth of a billionth or correspondingly huge ones, we need not only to understand them in a formal mathematical sense but also to reflect on their values and try to put them on some kind of comparative scale. When I was first told that the nucleus in the atom is so small that it can be compared to a speck of dust in the centre of the Royal Albert Hall, I got my first real insight into the vastness of the empty space inside an atom.

Electromagnetic waves

Radios or mobile phones are good examples of the kind of starting point we are discussing here. At first they might easily trigger straightforward questions — how do they work; how does one signal get picked out from all the others; how is the message carried? But, as we have seen above, they depend on the passage of electromagnetic waves from the transmitter to your particular apparatus; and trying to imagine this is far from straightforward. It seems almost impossible to conceive of these waves — all the TV stations, all the radio stations, and the emergency services and mobile phones, all in constant communication through the waves they put out into the space around them. Can we feel them? Why don't they interfere with each other? How do they get through buildings? So many questions arise because the waves cannot be seen or sensed by us and because the effects are so tiny.

Electromagnetic waves are vibrations of tiny electric and magnetic fields. Radio waves, which are one type of an electromagnetic wave, can vibrate up and down up to 300,000,000,000 times each second! Vibrations as rapid as this are impossible to imagine — for comparison, the typical vibrations of a piano or violin string are a mere 400 or 500 times per second. They travel at a speed of 300,000,000 metres each second. Such fast speeds are impossible to conceive — even an aeroplane only manages about 200 metres in a second. However, we can use these numbers to make comparisons. So, for example, the frequency of BBC Radio 4's waves is 92.4 million vibrations per second (called megahertz or MHz) whereas the frequency of BBC Radio 2's waves is 88.1 MHz. They are both extremely rapid, but nevertheless the precise difference between these two frequencies enables the tuner in your radio to separate out the two stations.

Radio and TV waves not only amaze us by their speed and frequency, they also confound us by their ability to permeate the space around us without in any way damaging or confusing each other! Transmissions from different broadcasters and countries all pass through the same space and coexist with TV waves, mobile phone signals and many other kinds of radio wave. In fact, the same is happening for the light waves emanating from every object we are able to see. They don't interfere with each other and confuse our perception of the visible world; they simply pass through one another without affecting each other. The sheer number of criss-crossing waves is indeed extreme but a simple analogy helps us visualise the process. The principle of waves passing through each other (a process called superposition) can be seen on any water surface where many waves interplay. In a harbour, for instance, waves from different boats criss-cross and reflect off the harbour walls; in a river, ripples from different disturbances — a paddling duck, a gasping fish, a bobbing piece of weed — run towards and then through each other, and pass on the other side quite unaffected. It's worth checking next time you are near an expanse of water! The passage of light waves and radio waves through the space around us works in much the same way; the signals simply pass unaltered through each other.

So, yes, the extreme properties of electromagnetic waves are difficult to grasp at first sight but, with time to reflect and to consider analogies, some appreciation of them becomes possible.

Biological processes

Quite different examples of extremes crop up in living systems. Cells are extraordinary examples of miniaturisation and are a regular topic for discussions based on everyday life. Nerve cells carry messages to and from the brain; blood cells transport oxygen from the lungs; stem cells, egg cells, sperm cells, white cells … a great range, all of importance in understanding the way our bodies work. Cells range from the visible (nerve cells can be many centimetres long) to the microscopic (cheek cells are roughly a twentieth of a millimetre — hundreds of times smaller). Even smaller are the components inside a cell that enable it to function. The most extreme example of biological miniaturisation that crops up regularly in discussion is DNA coiled up inside the nucleus of every cell.

In fact, DNA is one extremely long molecule composed of a scarcely believable number of repeating units, threaded together in a long, thin double helix. A useful image is of a kind of long, thin necklace, made of tiny links. In the case of DNA the number of "links" (or "bases" as they are called) on one chromosome (of which there are forty-six in a human cell) is approximately 220 million. Just how can a necklace with more than 200 million links be contained within the tiny nucleus of a tiny cell? The answer lies in another extreme fact: a DNA molecule is not only extremely long; it is also incredibly thin — just one or two billionths of a metre. At almost ten centimetres long its length is approximately 100 million times its width. Some necklace! The analogy can't be pushed much further, but by imagining a necklace coiled up in the palm of your hand you can get a sense of how a long thin chain can coil up into a small space. Now imagine a necklace thinner than a hundredth of a human hair, coiled up in your hand — that's how DNA gets packed into such a tiny nucleus.

The shape of the DNA molecule — extremely long and extremely thin — is just one of the extraordinary images we grapple with in addressing biological questions from everyday life. Many other features of our bodies at the microscopic level fascinate discussion groups. One is the sheer intricacy of the molecular mechanisms that underlie ordinary bodily processes. An example is the way information coded in DNA is transcribed to produce the enzymes and other proteins that run our bodily processes. In one discussion, following a description of the way various molecules physically interact with each other in sequential processes lasting tiny fractions of a second, one participant simply exclaimed, "It's mad, unbelievable — the way it all works on such a small scale." In a similar way the complexity of neuron circuits in the brain, the adaptability of molecules in the immune system, the molecular mechanism of muscle contraction are all examples of extraordinary processes at a microscopic level that are astonishing when first encountered.

Astronomic quantities

At the other end of the spectrum are the gigantic sizes associated with the universe as a whole. The Sun, planets, stars, space exploration, black holes and the Big Bang are equally popular topics for discussion as the smaller-scale objects here on Earth. They lead to contemplation of extremes of the opposite kind. The distances to stars, the time taken to travel to Mars, the number of stars in a galaxy, the age of the Earth, the Sun and the universe itself ... all involve unimaginably huge numbers. Just to get some sense of proportion, let us take distances: signals from the Moon take 1.3 seconds to reach us on Earth, light takes eight minutes to reach us from the Sun, but light from the nearest galaxy takes 2.3 million years. So again, quantities that are much too large to grasp alone can still be made good use of by making comparisons.

Equally extreme are temperatures in different parts of the universe. The vast expanse of empty space contains almost no matter and its temperature is considered to be only 3° above the absolute zero (the lowest possible temperature) approximately −270° Celsius. But from the colour of the light from stars it has been deduced that the temperature at their surface is between 3000° (for the cooler red ones) and 10000° (for the blueish-white hot ones). This is similar to the temperatures of red-hot or white-hot furnaces, or filament light bulbs, here on Earth. Calculation of the temperature in the interior of stars shows how very much hotter it is there: many *millions* of degrees — so hot that the very nuclei of the atoms that make up the star are able to fuse together. Indeed, it is this fusion reaction that generates the enormous quantities of heat that keep the star burning. This fusion reaction itself presents us with even more extreme values: the nuclei of atoms are held together so strongly that the fusion reaction generates extreme quantities of energy — tens of millions times more than typical chemical reactions, such as burning fossil fuels. Learning about the cosmos becomes a study of extreme values. The average mass of stars (2 000 000 000 000 000 000 000 000 000 000 kg) or their life expectancy (a few billion years) challenge our power of comprehension; likewise, the number of stars in a galaxy or galaxies in the universe, or the age of the universe itself.

The point of this brief excursion into extremes is not to strike you, the reader, dumb with awe — far from it. Its purpose is to bring out an important feature of the journey into science. Whatever the starting point, however mundane the initial question, we are led inevitably into a much larger range of values than we are ordinarily used to. Things can be enormously much larger or smaller, longer- or shorter-lived or more intricate or complex than the things we normally encounter. This can be a source of great wonderment and exhilaration, as many TV documentaries and web materials demonstrate, and a heightened sense of wonder is indeed one of the great rewards for enquiring into science. But an overemphasis on extreme values also runs the risk of intimidating or inhibiting enquiry. When confronted with a stream of unimaginably large (or small) numbers a person can easily switch off. The sheer unimaginability of extreme things can be stultifying — I can't imagine it, so I'll just ignore it.

Simply quantifying extreme values can also be off-putting in the ordinary arithmetical sense — after all, it's not easy to distinguish 10 000 000 from 100 000 000, even visually, let alone get a sense of their relative size. So in trying to engage with something of the wonder of science whilst at the same time strengthening one's grip on the actual size and complexity of things, dialogue is particularly helpful. Exchanges between those who communicate about scientific concepts and those who are grappling to understand them need to be two-way, so the difficulties anyone faces in comprehending unfamiliar things are brought to the surface. In the context of extreme values, this means going slowly (and repeatedly) over the magnitudes and complexities of things and searching for comparisons that enable the imagination to attach some kind of meaning to the strictly unimaginable.

Summary

This chapter has brought out a range of experiences people have when encountering scientific ideas seriously for the first time as adults. They are not necessarily the same kind of experiences they had as young people preparing for examinations at school or university. Drawing on hundreds of discussion sessions over many years, it is clear that a deeper understanding of science, not just the procedural knowledge needed to solve equations or recall facts, demands exploration of many kinds of issue that lie outside everyday experience. We need to become acquainted with the extreme sizes and scales of entities in the natural world. We need to extend our imagination to structures at both miniscule and gigantic scales. We require conceptual tools that organise and simplify vast arrays of information about substances and processes. The purpose of this chapter has been to expose some of these conceptual challenges and to offer a few thoughts about how to meet them.

Endnotes

i.   Villarreal, L.P. (2004). 'Are Viruses Alive?' *Scientific American*, 291(6), 77.

ii.   Singer, C. (1959). *A History of Scientific Ideas*. Oxford: Oxford University Press.

iii. Available at: http://www.lifesci.ucsb.edu/undergrad/pop_up/bacterium/bacterium.html.

iv.   Available at: http://www.newscientist.com/gallery/mg20327200-virus-killer/2.

v.   Available at: http://upload.wikimedia.org/wikipedia/commons/1/1d/Poliovirus_binding_receptor_1DGI.png?uselang=en-gb.

vi.   Available at: http://www.physik.uni-augsburg.de/exp6/imagegallery/afmimages/afmimages_e.shtml.

vii. Available at: http://www.nei.nih.gov/eyeonnei/snapshot/archive/051809.asp.

viii. Available at: http://upload.wikimedia.org/wikipedia/commons/0/07/Water_molecule.svg.

ix.   Available at: http://upload.wikimedia.org/wikipedia/commons/thumb/1/1b/Ascorbic-acid-3D-vdW.png/574px-Ascorbic-acid-3D-vdW.png.

x.   Available at: http://img.readtiger.com/wkp/en/Polyethylene-3D-vdW.png.

xi.   Available at: http://en.wikipedia.org/wiki/Grain_boundary.

xii. Available at: http://www.aps.org/meetings/march/vpr/2009/imagegallery/silico-natoms.cfm.

xiii. Available at: http://commons.wikimedia.org/wiki/File:Atom_diagram.png.

xiv. Available at: http://scienceray.com/biology/proteins/.

xv.   Available at: http://www.youtube.com/watch?v=5C5_dOEyAfk.

xvi. Isaac Newton, Letters to Bentley, 1692/3. Quoted in Wikipedia at http://en.wikipedia.org/wiki/Action_at_a_distance_(physics).

Chapter 7

Following Up Your Curiosity

In earlier chapters we considered the kinds of things that arouse our curiosity about the natural world: questions we have always wanted to ask triggered by things we notice in the world around us. Curiosity may lead you to pose a question in your head, perhaps express it to a friend, but all too often it just remains hanging in the air, effectively a rhetorical question. On a good day, however, you may be stirred to pursue a query a little further, perhaps by getting hold of a popular science book or maybe visiting a museum. In this chapter we look at some of the challenges that arise when you decide to follow through by taking your questions more seriously, and suggest some ways of meeting these challenges.

Language

Whether you pick up a book or magazine, go to a talk or try recalling a science lesson at school, difficulties often arise in trying to understand some aspects of the language used. This can happen even when a book or talk is said to be accessible to non-specialists. The problem is very common and is even experienced by people trained in science. There are many reasons for it, some legitimate, some not. We touched on this issue in Chapter 3 and explore it in greater detail here.

On the positive side, technical language develops as a kind of shorthand in all professional communities. It helps internal communications by speeding up reference to commonly occurring things. So, to some extent, it is worth engaging with this by grappling with some new words: those that occur regularly — like molecule, cell or field — where a long-winded phrase would otherwise be necessary. Other words can of course be looked up as the need arises, but this can occur so often that reading is constantly interrupted. It can also it lead to frustrating iterations in which one definition is given in terms of other unfamiliar words. In this situation it may simply be better to drop it and look for a different book rather than give up the whole subject in despair. To illustrate the point here are two explanations of the word "work", as used in physics. The first is from a popular website, the other from a scientist who writes well for the layperson.

1. In physics, a force is said to do work when it acts on a body, and there is a displacement of the point of application in the direction of the force.
2. Work is motion against an opposing force.

Sometimes, technical language is used because the intended readership is technical; but sometimes it is just used thoughtlessly. The challenge for communicators who are trying to reach outside the professional community is to dispense with unnecessary technical vocabulary and find alternatives. Some manage to do this and some don't; the reader has to browse and select carefully.

Vocabulary presents an additional, rather particular problem in science. Words that have quite general meanings in everyday life may have been hijacked for more specific purposes in science — or vice versa. As one discussion group discovered, this has even happened for the word "water" (see the example below).

Ambiguous Language

An example from a packet of tea

When exploring why a tea packet advised that "re-boiled water should not be used because it has lost its oxygen", it gradually became clear that the simple word "water" has more than one connotation — a point that is extremely easy for a scientist to overlook. In everyday language, it refers to the liquid that comes out of a tap or a bottle and is used to make a cup of tea. In science it can also be used to refer specifically to the H_2O molecule. Of course, the familiar liquid from the tap is overwhelmingly composed of H_2O molecules, but it also includes many other molecules in small quantities — the impurities that it picks up from the air and earth through which it has travelled. This ambiguity in language was the cause of a fundamental misunderstanding about where the oxygen comes from that gets lost when water is re-boiled — from the H_2O molecule itself or the free oxygen dissolved in it.

Other words such as energy, power, charge and field, taken from everyday life, have precisely defined meanings in science that may not align well with their everyday connotations.[i] One person illustrated this point rather nicely when she commented in a discussion about magnetism: "You can't just throw these words around. When I first heard about magnetic fields, the image that came to mind was of something green and grassy."

Vocabulary is not the only language problem; some writing about science is simply too dense. In one form of this, too many ideas are packed into a single sentence. If these ideas are unfamiliar you barely have time to grasp the meaning of one before the next is upon you; the pace of sentences needs to be slower. In a slightly different form of dense writing the steps in a line of argument are simply too far apart; in effect the writer misses out intermediate steps, perhaps because they are so obvious to the writer. When you are new to a subject it's hard to know where or whether any steps are missing; if you cannot see the link from one step to the next in an argument it may reflect more on the quality of writing than on your ability. After giving it a few tries it may be best to trust your judgement and change books rather than blame yourself for failing to understand.

So for non-scientists, it's a matter of selecting carefully what to read. My own experience is that incomprehensible writing is relatively common, so you need to dip into any book cautiously before deciding whether to read it thoroughly. It's not so easy to judge when you have little personal experience of science, but if in doubt I wouldn't be afraid to reject a book if you are failing to make headway. The risk of missing out on a good book is outweighed by the far graver risk of feeling defeated by science once again, having your fears reinforced and giving up on the whole project! With this in mind it may be more economical to borrow from a library first to check out an author. But there is good news too. The rise in popular science in recent times seems to be encouraging greater numbers of professional writers into the field, resulting in a growing range of highly readable books on scientific topics. So in many areas of science there is now a reasonable choice.

Cognitive Challenges

However important clear language is it would not be wise to pin all the problems of learning science on authors and broadcasters alone. Even when communications are at their best, some issues in science are inherently difficult to comprehend. An important review of science education[ii] found: "[There is] very strong empirical evidence that some of the fundamental concepts on which scientific understanding is built are commonly misunderstood by learners, and that there are patterns in the difficulties that they experience."

There are different views about whether science is intrinsically harder than other branches of knowledge. Personally, I don't think it is worth worrying too much about this question — philosophy also seems to me to have some tricky ideas to grasp, as do law and economics and branches of sociology, political science and cultural studies, to name just a few. Some studies of school science suggest that students find it harder than other examination subjects. A UK study in 2008[iii] stated: "There is a clear indication that the STEM (Science, Technology, Engineering and Mathematics) subjects are generally harder than other subjects."[iv] For adults trying to gain some insight into scientific ideas, I doubt if cognitive difficulty is the most important barrier: there are plenty of more pressing practical ones. Nevertheless the experience of working with discussion groups shows even when such barriers are down, some topics present real thinking challenges. Paradoxically, the non-threatening settings of informal science discussions means these challenges are often relished, with that particular sense of satisfaction that comes from finally grasping a concept that has eluded you for decades.

The discussion process described in Chapter 1 involves several distinct stages in the understanding of new ideas, each of which has its challenges. The first, trying to articulate what you already know or believe, is not as easy as it seems, especially if you feel uncertain about your knowledge of the matter in hand. Second, allowing challenges to what you previously believed is something we all find difficult, in many areas of life; letting go of old perceptions in the light of new evidence is even harder. Finally, incorporating new understanding into our ways of thinking, making use of new insight, takes quite some extra effort. Despite all of this we can see by simply watching children in everyday situations that it is all possible — we just need to recapture the spirit of enquiry.

Let us look at each of these challenges in order. To articulate what you already think or know about a topic calls for a degree of confidence about saying what you think and at the same time forbearance on the part of those around you — they need to hold back before pouncing on any error they may perceive in your reasoning. Initially, weak ideas as well as strong ones need to be allowed to surface and then subjected to reason and argument. In this way flaws can be revealed and a collective move made towards better reasoning. An illustration of this process in a discussion group is given below.

The free exchange of thoughts helps to reveal which beliefs are actually in play before the science cuts in. Acknowledging and examining these are important preliminaries to eventually changing them, as the example on the next page shows.

As discussed in Chapter 5 the issue of "charge" reveals a further and deeper issue about how different people deal with scientific explanation. In a discussion about electricity one person said bluntly, "Every time you mention the word 'charge' I lose the thread and stop listening. I just don't have any idea what it means." Another person, however, was content to say, "I can't explain it; I just get on and use the word." My experience is that both these

Expressing your First Thoughts

The physics of floating

Discussion once started with an observation that icebergs float mostly below the surface whereas human beings do so mostly above it. Participants were encouraged to express their first thoughts about this: one remembered that water expands in a bottle put in the freezer, then may burst; another asked if it got heavier, others rejected the idea — how could it? The volume must get larger, suggested another, not certain what volume really meant. The idea of wood floating on water was introduced — does this depend on its weight? Does a log float in the same way that a twig does? What about a waterlogged log? Free discussion plus input from the tutor enabled the meaning of density, weight and volume to be grasped so that the physics of floating could be understood. The anomalous structure of ice was introduced to explain why it floats in water; discussion was finally exhausted as someone declared, "Water is just weird! You'd expect stuff to get less dense when it goes from solid to liquid." The complicated behaviour of ice could be explained only because the basic ideas of weight, volume and density had been thoroughly tussled over first.

Sorting out the Flawed Ideas

The magic of radio

A discussion once started with the topic of sound as heard on the radio, TV or phone. How did the sounds reach us? After introducing the idea of sound as a wave the first idea put forward tentatively was that the sound wave itself travelled through the air to a radio or TV receiver. Once it was realised that this could not be the case — we would be inundated by sounds from everywhere — the idea of a sound-wave's profile being impressed upon an electromagnetic wave that carried it was introduced. This in turn gave rise to exchanges about the meanings of electromagnetism — "I assume it's a combination of electrical and magnetic attraction?" one person queried. Another suggested: "Is magnetic attraction a pressure, a pulling thing?" A third asked: "Doesn't there have to be positive and negative charge?" A fourth raised a more fundamental issue: "What is charge anyway?" These initial thoughts included a mixture of valid and false points which helped clarify where the conceptual difficulties lay. It was only after these areas of confusion about electrical and magnetic effects had been surfaced that a scientific explanation of the nature of radio transmission could be usefully put.

ways of dealing with the unfamiliar have their place: probing the deeper meaning of words on the one hand, and getting used to using them in practice, on the other. Infants seem to pick up words and use them freely well before they have grasped all the nuances of their meaning. Sometimes simply working with unfamiliar language seems to make it easier to accept and ultimately to adopt. You get to understand charge as something that is stored in a battery and gradually drains out when you use it to light a torch or run a mobile phone. On the other hand, explaining more deeply that charge is a human construct invented to explain the behaviour of substances that can attract or repel each other (like a comb and a lock of hair), opens up a deeper understanding of scientific theory and how it emerges.

Once our initial thoughts about a scientific question have been bounced around and a scientific explanation provided, we may find some of our prior beliefs seriously jolted. With the prospect of our long-held understanding being forced to change, it is likely we will put up some understandable resistance, at least to begin with. In one discussion about the internal structure of atoms a participant reacted to the extraordinary (and unexpected) results of Rutherford's famous 1911 experiment[v] by protesting: "It seems impossible that an atom is almost entirely empty space; everything feels solid!"

Sometimes metaphors can seem more of a hindrance than a help. In a discussion about the arrangement of electrons inside an atom, the commonly used metaphor of shells to describe the arrangement of atoms was criticised as "a confusing image — it implies that there is some physical structure like an eggshell there in the atom". It caused someone to ask, "Why don't the electrons crash into one another in a shell?"

In a completely different discussion about clouds, a fierce argument once broke out about whether clouds have edges. This in turn proved a trigger for a deeper discussion about distributions of density and probability — the way in which effects can tail off, getting as weak as you like, without ever theoretically reaching zero. This proved a key point in grasping how the intensity of light and the effects of gravity or magnetism diminish as you move away from the source, becoming as small as can be imagined but never absolutely zero.

Another kind of cognitive jolt can come from re-examining something vaguely grasped at school. The nature of light and seeing are examples of this. After a discussion about sources of light, such as the Sun or domestic bulbs and the way in which light from them is reflected off objects into our eyes, one person commented, "Sometimes we never really take on board simple scientific things we have been told. It's quite a shock to realise we only see objects because light has bounced off them into our eyes."

The nature of the vacuum is another example of a familiar concept that triggers deeper thinking. In a discussion about ice, it had been explained that the H_2O atoms were held together in a lattice with empty space in between. One person responded: "It freaks you out that there is just a vacuum between molecules in ice (or anything else). Still, air is 99% empty space I suppose. Is it really a vacuum or just that we don't know?"

The sheer emptiness of outer space is also a shocking thing to try to grasp. The idea that light which travels to us from distant stars without having to latch on to any substance during its journey is hard to accept; in our everyday experience waves and vibrations exist in some kind of medium, like the water of a pond, the skin of a drum. No wonder it is hard for ordinary people to grasp the passage of an electromagnetic wave through the vacuum of space, when a mere hundred years ago a scientific periodical wrote of the invisible ether, believed at the time to fill all space: [vi]

> At the present time there are three theories [of the ether] ... the third denies its existence altogether ... we see re-appearing the mysterious and rather terrifying notion of the absolute "nothingness" of outer space which one imagined to be successfully abolished by the introduction of the ether.

In the case of biological matter other kinds of conceptual challenge exist. A key one is the temptation to attribute human characteristics to non-human things, in particular to impute purpose rather than accepting that things can simply occur without purpose. In a typical example that arose in a discussion of genetics it was explained that "the DNA in a chromosome unravels so that each particular gene is exposed and can be read off". Discussion developed around the idea that the DNA did not "mean to unravel, or wish to"; there was no sense of purpose. Causes of the unravelling can be explained at various levels — immediate causes such as the presence of an "unravelling" enzyme and deeper causes such as the electrical attraction of some atoms on the enzyme to other atoms on the DNA. The challenge is to understand the limitations of cause-and-effect explanation and to accept the less demanding (and perhaps less satisfying) level of explanation: this is what is observed to happen.

Perhaps the most counter-intuitive area of biological explanation is the theory of evolution by natural selection. The idea that random variation in individuals can confer a greater chance of survival, and hence of reproduction, is not so difficult to grasp. But the idea that complex anatomical and behavioural characteristics have been arrived at solely by this means, is difficult to believe. That organs like the brain have emerged from chance variations in anatomy and that elaborate social behaviour in bees or baboons has simply evolved seems impossible. Perhaps the underlying conceptual difficulty is in imagining the extremely long periods of time over which these adaptations took place. Where evolutionary change can be illustrated over much shorter periods the concept is more readily accepted; an example is the loss of light-coloured moths and increase in dark-coloured ones in recent centuries as industrial pollution darkened the trees on which they lived.

In some specific areas of life the Western tradition of scientific explanation clashes directly with alternative systems of belief. Eastern medical traditions, for example, explain health conditions in a quite different way from the West and prescribe different kinds of treatment. A typical discussion involving alternative or complementary forms of medicine focused on "energy tracks or meridians that deliver energy to parts of the body". Energy was described as "flowing through a particular organ at a particular time of day. You treat according to the time of day". Debates of this kind often compare Western and Eastern approaches and contrast the treatment of symptoms with responding to deeper causes. Similar debates arise in discussing psychological therapies and other ways of addressing the sense of wellbeing. Given the popularity and variety of such approaches and the widely read literature about them it is important for science discussions to embrace these varying perspectives; they provide a good opportunity to debate the nature of scientific evidence and its limitations in areas of life that are important to people, such as fitness and mental health.

Social Challenges

Most members of the discussion groups have participated consistently over many years (in some cases, more than twelve years). One factor in this persistence is the social setting in which the learning takes place; the informality enables people to talk about science freely with one another as well as to interact with a tutor. The importance of an appropriate social setting is also reflected in studies of science at school. A review of educational research[vii] by the UK National Foundation for Educational Research commends "the notion of developing a classroom climate in which pupils are willing to discuss their ideas and are not afraid of being wrong" and goes on to say that "research shows willingness to discuss ideas in a non-threatening environment is crucial to finding out and addressing pupil misconceptions."

For participants to avoid feeling threatened when they offer their preliminary thoughts, others in the group have to be careful to avoid put-downs. It is extremely easy for a person who believes they know something to silence a person effectively who believes they don't. Frequently this leads to clamming up by the very people whose perceptions need to be brought out. This diminishes the overall learning possibilities because it not only makes one person feel slighted and less likely to participate fully in discussion but also blocks off insight into how concepts are perceived and hence how any misconceptions might best be addressed.

It can be helpful to be aware of gender issues in science groups. Research on the issue with young people at school shows that only a minority of girls engage with the physical sciences, though the reasons for this are not easy to establish given that many factors affect students choice of subjects: parental aspirations, sense of own ability, quality of teaching and sense of identity.[viii] However, there is evidence that girls and boys have quite different favourite topics in science: explosions, weapons and space for boys; dreams, illnesses and the human body, for girls (to paraphrase the results of an international study[ix]). Experience in the author's discussion groups is of a much less clear-cut divide for adult learners. Cosmology, atomic physics and molecular chemistry seem to fascinate both sexes, similarly with biological issues.

A more significant gender issue in discussion groups seems to be the manner in which scientific issues are approached. By observation there seems to be stronger interest amongst women in entertaining uncertainty, posing questions and exploring rather than directly answering questions. It would be interesting to know if research on these kinds of preference amongst adult learners has been undertaken. From the point of view of successful group interaction it seems to me important that men and women understand that the way they wish to approach topics may differ and that both need to be accommodated. In open-ended explorations it is important that excessive exchange of factual information doesn't crowd out the expression of uncertainties. The tutor plays an important role in maintaining a balance between approaches and in ensuring discussion remains sufficiently open-ended.

Remembering

"I don't remember anything I learned in science at school" is a common cry from people returning to science in adulthood. In fact it is not usually strictly true. It is interesting how, with a bit of provocation, odd memories from school do come back: semi-permeable membranes, blue crystals of copper sulphate, mirrors that turn your face upside down. However, when it comes to fundamental ideas about genes, cells, forces or chemical reactions, for example, most people in the author's groups recall very little conceptual understanding from earlier learning. "Volts over amps equals ohms!" shouted out one participant in a group discussion, suddenly recalling an ancient phrase. "I remembered it for the exams but never understood it, then or now." Paradoxically, learning at school, at least at secondary level, is strongly associated with memorisation, not least for written examinations. Because of the traditionally strong connection between science and factual memory it seems important to consider what role remembering should or does play in adult science. Do people remember the science they learn in informal settings, with no examination or syllabus to drive them? No formal assessment has been made of this in the author's discussion groups, but some trends are revealed in interviews with participants.

On the whole, factual details about scientific phenomena do not seem to be retained for long after a discussion. As one group member once began: "I've read this interesting book about relativity. It says there are three main things to remember: invariance; causality and ... I've forgotten the last one." When a topic such as electrical circuits or cellular components is revisited after some time, most of the detail — naming of parts, precise functions, etc. — are not recalled. To a limited extent this may seem disappointing to participants: "We've been over this before haven't we? I can't remember what was said." It is true that if the learning aim had been to acquire factual or procedural knowledge, an informal, unassessed way of learning would probably not be advisable. I suspect we would all prefer that our airline pilots and medical practitioners are able to recall vital facts when needed!

But different criteria are appropriate for informal learning. Affective outcomes are also important for many who make the effort to learn through open-ended discussion. A growing sense of confidence about joining in scientific conversation or reading a science-based book or following scientific and technological aspects of current affairs was the main benefit that people brought out in a focus group discussion on what participants felt they had gained. A second consideration was, to quote one participant, "Feeling you have connections to the other half of the world." Or in the words of another: "You have connections with things in the world, as in other areas such as politics." It also helps people feel on a par with others: "You're more confident to ask questions. You can see that a person's apparent knowledge can be just superficial. A person may appear scientifically confident but be unable to answer questions." The ability to apply one's scientific knowledge more directly did not seem to be as important: "You don't use it in everyday life but you bring daily life to the group; it expands your thinking, you can connect up."

So remembering facts does not seem to be such a valued outcome for the participants. From the tutor's point of view, it is clear that by revisiting fundamental concepts such as atoms, molecules, elements, cells and genes, over and over again in different contexts, a broad conceptual understanding does grow with time and previous, less scientific conceptions do get modified gradually. It may be this growing conceptual grasp that underlies the sense of confidence that participants describe.

Memory isn't all it's cracked up to be, either! It comes as quite a surprise when people meet up with practising scientists and discover that they too don't remember everything they learned at school or university and they too may be quite ignorant of topics outside their own specialist area. Explaining this reality can be helpful to non-specialists. The body of factual knowledge in any area of science is far too big to retain; when a scientist needs to know something specific they will usually look it up. A similar approach is also central to informal learning for adults; both the tutor and participants are free to pose questions and pursue lines of enquiry beyond their fingertip knowledge by seeking the factual detail they need outside of the discussion session. Of course, the precise way in which this is enabled is changing extremely fast. Discussion groups that once relied on book reading between sessions now use the internet for rapid access to resources. More recently, tablets and smartphones are used to answer points on the spot. These help establish the meaning of unfamiliar terms and new concepts and provide images, animations and video clips that illustrate things graphically. The latter are particularly helpful for understanding complex entities such as the structure of molecules or regions of the brain and dynamic concepts such as the motion of planet or growth of cells.

So for the person wishing to learn more about science but wondering whether it will all be forgotten and therefore a waste of time, the message is: think about what you want from the experience. If it is reliable, retained knowledge about a specific topic or procedure, it is probably best to join a course about that topic, with its schedule of assignments and assessments, to ensure knowledge is absorbed and retained. If a non-formal setting is preferred, stand by for a gradually developing insight into broadly based concepts and a sense that you can begin to connect with conversations and media coverage of scientific issues.

Accumulating Knowledge

One of the benefits of learning informally is that it is easier to stick with it for longer — exam stress is absent and attendance is voluntary, so it's easier to fit round a busy adult life. As a result people can stay with the subject for many years, much as for a book group or study circle. With such longevity comes the opportunity to revisit topics time and again and build knowledge gradually. This so-called "spiral" approach to a curriculum means that concepts are encountered repeatedly, helping to embed the unfamiliar. The spiral image implies a degree of forward movement as well as circling round. In relation to learning, each repeated encounter with a topic arises from a different starting point in everyday life and these varying contexts enable a concept to be looked at from several different angles; the long-term effect is to deepen understanding. With time, links begin to be seen between one topic and another. Electrons crop up as part of the inner structure of atoms, but they also appear in electric currents and again in the action of acids. At first, DNA may appear in connection with crime investigation then again as part of the machinery of reproduction or the mechanism of evolution. Various strands of discussion begin to connect up. As one group member once put it: "It is good to have somewhere to ask all these questions and see the connections between bits of knowledge and experience; we will be here till we are 70."

In this way general points are developed from particular instances. Aspects of everyday life become the inspiration for grappling with theory. And as knowledge begins to accumulate you begin to understand some of the categories in science — it's not just a stream of arbitrary facts. As connections are made and categories develop, one of the most empowering aspects of learning becomes possible: the opportunity to apply a general concept, developed in one context, to a different situation (see example on the next page).

Of course, it is no simple matter to apply knowledge acquired in one context to a new one; this is a fundamental problem for teaching and learning even in formal, structured learning environments. There is always a risk

Using Knowledge from One Context in Another

Neurophysiology and child development

As an example, one discussion began with a discussion of brain plasticity (its tendency to be shaped in response to the environment), based on a book one participant had read by a psychotherapist writing about interactions between babies and parents. This topic was then linked to an earlier anatomical discussion about physical areas of the brain (Broca's area and Wernicke's area) and led on to an input from another participant who was aware that babies growing up in bilingual environments had the ability to detect a third language being spoken. Connections were thus made between knowledge held by three people with different areas of experience, which drew on research in several distinct disciplines: neuroscience, psychotherapy and linguistics.

that, without guidance, previous knowledge will be misapplied in a novel context. But the reality is that we all risk misapplying our partial knowledge to some extent in everyday life, just to get by! The accumulation of a repertoire of scientific concepts through sustained exploration means that new ideas can at least be approached more intelligently than otherwise, with less risk of misinterpretation.

Summary

In this chapter we have considered what follows after curiosity has been aroused, questions posed and an exploration made of the relevant science. It is, of course, all too easy for us to idly pose a question ("Why is raining so much these days?") and then let it go with a wistful sigh. To muster the strength to pursue it in discussion or through reading is itself something of an achievement. But as this chapter has indicated, even if you do so, a number of difficulties lie ahead and it is as well to be prepared for them so they may be overcome. The most obvious ones we have considered are related to language: writing that may be impenetrable, ambiguous or incomplete. Others are cognitive, concerning inherent difficulties in thinking through complex ideas; still others are social, involving the risk of put-downs or feelings of inadequacy in revealing what you do or don't already know. None of these difficulties need be a deterrent to exploring scientific ideas. Instead, by preparing for them and adopting some of the strategies suggested for working round them, the way is open to gaining confidence about engaging with the many scientific debates that fill our newspapers and TV screens.

Endnotes

i. For a fascinating exploration of the role of language in science, see: Sutton, C. (1992). *Words, Science and Learning*. Buckingham: Open University Press.

ii. TLRP (2011). *Science Education in Schools: Issues, Evidence and Proposals. A Commentary by the Teaching and Learning Research Programme*, p. 8. Available at: www.tlrp.org/pub/documents/TLRP_Science_Commentary_FINAL.pdf.

iii. Coe, R., Searle, J., Barmby, P., Jones, K. and Higgins, S. (2008). *Relative Difficulty of Examinations in Different Subjects*. Durham, UK: CEM Centre. (Also see: http://www.cemcentre.org/documents/news/SCORE2008report.pdf.)

iv. An experiment in which particles were shot into the atoms of a piece of gold foil and were seen to bounce back unexpectedly, "as if you fired a 15-inch shell at a piece of tissue paper and it came back and hit you", as Rutherford expressed it.

v. Anon. (2014). *Scientific American*, 310(2), 65.

vi. Hodgson, C. and Pyle, K. (2010). *A Literature Review of Assessment for Learning in Science*. Slough, UK: NFER.

vii. Osborne, J. and Dillon, J. (2008). 'Science education in Europe: critical reflections. A report to the Nuffield Foundation.' Nuffield Foundation, London, p. 16. Available at: http://www.nuffieldfoundation.org/science-education-europe.

viii. *Ibid*.

Chapter 8

Taking Things Further

Having addressed some of the challenges in taking your curiosity about the natural world seriously, we now consider the next step: how to extend understanding beyond your initial questions. We look at the kinds of resource available to help dig deeper and to develop a broad understanding of a topic area. First we tackle the practical issues, then some of the intellectual ones.

Resources

At first sight, getting into science may not seem an easy thing to do on one's own, as an adult. Where once teachers and courses provided an external structure for learning, now you are on your own, with whatever resources can be found at home or nearby. Scientific facilities for the layperson seem rather meagre out in the community: where are the public laboratories to compare with the art galleries, theatres and cinemas that cater for the arts? On closer examination, however, a wider range of science-based resources than you might expect does actually exist in the community. With some imagination, and a little investigation, some of these can be commandeered to help in the quest to take learning a little further. We begin by considering the simplest of these: the library or bookshop.

Books and magazines

Publishing in popular science has simply exploded in recent years. Where once there might have been a book or two on a subject, written by a scientist keen on communicating to the public, now there is often a range, with an increasing number written in highly accessible language by professional writers. The issue becomes not so much a problem of finding a book, more one of choosing and affording one. The costs can be minimised either by ordering through a public library or by forming a discussion group and sharing books. Choosing is more complicated — the challenge is to find something that is both good science and understandable. In my experience it is better to flick through several in a bookshop or library with these two tests in mind, trusting your own judgement. With popular science books it's important to try out a few paragraphs at different points in the book to see if it seems readable throughout, not just at the beginning. For people whose interests are wide-ranging it is better to be reasonably open-minded about which topic to choose (evolution, earth science, cosmology, for example) but highly selective about the quality of writing. Some well-written books, such as Dava Sobel's *Longitude* and *Galileo's Daughter*, focus on very specific topics but open up broad areas of science, linked to other aspects of life. Others, such as Bill Bryson's *Short History of Everything*, make highly attractive reading about a wide range of scientific topics that interest the author. More systematic coverage is offered by books by specialist science writers, such as John Gribbin with his *Science: A History* or Matt Ridley with *Genome* and *The Origins of Virtue*. A large number of popular science books today are written by scientists who have developed a skill in writing. Many of these are well written but it is important to check that the level of prior knowledge assumed and the use of technical vocabulary and mathematics are suitable for you. A list of accessible books that have been read and enjoyed by people in science groups is available on the *Getting to Grips with Science* website (http://gtgwithscience.com) with comments from readers.

Magazines and periodicals abound in science, as in most fields. Some are specialist, aimed at enthusiasts in, for example, wildlife, engineering or archaeology; these will include some scientific or technological material as a matter of course. Others put science centre stage. Some, such as *New Scientist*, aim to keep abreast of breaking news in science and are organised more or less as a weekly magazine. Some articles require a level of technical background, others don't. A different kind of periodical appears less frequently and provides longer pieces, following roughly contemporary research. The prime example, *Scientific American*, is edited scrupulously to ensure that any background technical information required to understand an article is summarised briefly at the beginning and technical terms are explained as they arise. It is famed for its graphics, which are used extensively to explain complex phenomena.

After the purchase (or loan) comes the reckoning — will I get through this book and be the wiser for it? Some, of course, work on a narrative basis: either you get into the swing of the story and follow it through to the end or you decide to drop it at some point. No special considerations apply to stories of this kind just because they are science-based. In others, however, understanding builds gradually through a sequence of chapters. My experience is that these are of two kinds and distinguishing between them is crucial for a non-specialist trying to get the gist of a new topic. In the first kind, each chapter builds critically upon the points made in previous ones. Essentially, like a textbook, it is impossible to make sense of later chapters if the earlier ones have not been reasonably well grasped; more advanced concepts are described in terms of preliminary ones described earlier. In this case you either acquire understanding as you proceed and reach the end or you decide to finish at some intermediate point. In my view, part-reading of this kind is a perfectly reasonable strategy, leaving you better informed than before and open to continuing with the book later on when your background knowledge has grown.

In the second type the dependence of each chapter on the previous ones is less critical. Although the book drives towards higher levels of understanding, parts of the build-up can be missed out safely. Some authors even go so far as to indicate where a chapter can be missed without loss of overall understanding, perhaps a mathematical diversion that is not crucial to the argument. Clearly it is better to read such books through to the end, skipping where necessary, rather than giving up altogether. A few books are structured even more flexibly so that chapters can be dipped into as desired because each chapter is conceptually free-standing. A general point about good reading practice is particularly helpful for science: when dipping into a book it helps to have a question in mind; in other words, take control by interrogating the text. Thus, when trying to work out what methylation has to do with epigenetics, it may sometimes be better to scan a text for this word and read around it rather than to plough through a lengthy text and risk losing heart.

The web

The web has become a source of instant information beyond the dreams of teachers and learners years ago; the diameter of the Earth, the wingspan of a pterodactyl, the 57th element in the periodic table — all available at the click of a button. Not only is such information available instantly, it can also be cross-checked against several sources at the same time, a huge advance over encyclopaedias and textbooks in libraries remote from home and work. Of course, with no overall quality control there is plenty of room for misinformation, and the surfer has to be confident about the sources chosen.

Instant access to information has deeper consequences, which work in favour of the layperson developing an interest in scientific ideas. No longer is it essential to memorise so many basic facts to ground your understanding, instead the emphasis can now be on developing conceptual understanding. Factual information can be checked for accuracy as understanding grows and can be recalled as and when it is required.

But of course there's more to the web than words, images too are a great help in trying to get to grips with scientific ideas. No amount of text can compete with the sheer elegance of the periodic table as a diagram or the majesty of the aurora borealis on a video clip. Photos of molecule, even atoms, taken with the most advanced electron microscopes bring the invisible to life; animations of DNA replicating or planets orbiting show clearly what is happening in complex time-dependent processes.

Several types of source on the web have proved invaluable in helping discussion groups pursue their questions. Sometimes it is just simple information that is required — the size of an atom or date of the Jurassic period — and encyclopaedias can supply this. On other occasions it is more extended explanation that is sought. Typically this occurs for topics where neither the tutor nor participants in a discussion group has enough background information but the issue is of topical interest. An example was the issue of dark matter, a relatively recent research topic and source of occasional news stories; another was the discovery of the Higgs boson. In these situations the web provides the essential information and the tutor's job is to select the best web sources and to help interpret what they say.

The web also gives access to structured learning materials. Some come from national institutions such as the BBC or a university, others from organisations with particular interests, such as an agency, lobby group or commercial firm, and a few are from individuals turning their hand to teaching. Clearly, it is important to know which of these types is being looked at, in order to work out whether it can be relied up for accuracy and clarity. The Open University[i] is a particularly useful resource with its links to radio and TV broadcast in addition to its written material.

When a specific question is asked via the internet — "What is meant by energy?", for example — the responses can be unhelpful. In many cases the explanations are too brief and simply raise further questions about the meanings of the words in the answer. Sometimes they simply lead into forums where people exchange thoughts about the question rather than offer an authoritative response. Occasionally they are misleading or just plain wrong. So, once again the trick is to look at the web address and try to judge how trustworthy the source is likely to be; then check out the credentials of the organisation behind the website. Although the web is sometimes criticised for its openness to any source without regard to quality, it is striking just how many prestigious institutions in science (and engineering and medicine) have a significant presence and how many resources of great value to the public are available via their websites. Professional institutes, government bodies, research centres, universities, museums, environmental and other scientific agencies, special interest networks — all contribute useful and readable material via the web. Some examples are given in the list below:

- BBC Science
- British Science Association
- Society of Biology
- Institute of Civil Engineers
- Institute of Electrical Engineers
- Institute of Mechanical Engineers
- Institute of Physics
- Natural History Museum, London
- Open University
- Royal Institution
- Royal Society
- Royal Society of Chemistry
- Science Museum, London
- Scientific American

Lectures and talks

The written word and printed image are inevitably key resources for learning about science; they are available at the drop of a hat, in every place and, in particular, in your very own home. From time to time, however, the word of the expert, delivered live, perhaps open to questioning, can bring a subject to life and perk up curiosity. Unfortunately, these won't be available in every locality but can often be found at a university, either directly from the science departments or through so-called "extra-mural" provision. Other local institutions may also offer talks: science-based societies in, for example, astronomy or nature conservation; adult learning organisations such as the Women's Institutes, University of the Third Age and local museums are just some of the possibilities. A report on opportunities for adults learning about science[ii] was published by the National Institute of Adult and Continuing Education in 2001, which gives several examples of such opportunities.

The national institutions for scientists also offer a wide range of talks and demonstrations for the public and scientists alike. For example, the Institute of Physics[iii] has local branches across the country offering lectures and demonstrations on topics such as lasers, crystallography, gastronomy and quantum mechanics. The Royal Society of Chemistry[iv] organises events all over the country on topics ranging from ocean debris to the chemistry of flavour. The Society of Biology[v] has similar provision on topics such as prostate cancer, nature reserves and toxicology. The Royal Institution in London, famous for its televised Christmas lectures for young people, also offers a fascinating range of talks for the general public.

Such talks are usually given by scientists and ex-scientists themselves rather than by professional science communicators. It is wise therefore to remember in advance that sometimes technical language may creep in or assumptions made unwittingly about levels of prior knowledge. In this way allowance can be made for potential communication difficulties and preparation made to ask questions if permitted. The experience of discussion groups in listening to scientists is very positive. With the increasing emphasis on engaging with the public, many scientists are developing skills in talking to non-specialists. It is also the experience of the groups that a non-didactic approach works best in which the layperson is able to pose questions and ask for clarification as discussion proceeds.

Talks are, of course, no longer restricted to live events. The growth of podcasting and videostreaming means that high-quality expert talks are available anywhere via the internet. These often combine serious science with an entertaining style of delivery and are highly suitable for the curious non-specialist. Some websites that provide high-quality talks online are intro-duced below.

Online Talks

Little Atoms is a weekly radio series, and produces podcasts by experts on a wide range of scientific topics that can be downloaded for free.[vi] Examples of titles include: "What It's Like to Be a Bird", "The Self Illusion" and "The Epigenetics Revolution". It describes itself as "a talk show about ideas and culture, broadcast every Saturday at 10 am (and repeated every Sunday at 7am) on Resonance 104.4 FM". The podcasts are available on iTunes and at http://feeds2.feedburner.com/littleatomspodcast.

TED Talks[vii] (Technology, Entertainment, Design) are video talks, some of which cover topics in science and technology, often in relation to wider social issues. Examples include: "Why Do We Sleep?", "The Levitating Superconductor" and "How a Fly Flies". Given by experts before an audience they are usually entertaining and can be downloaded free of charge. The website explains that:

> TED conferences bring together the world's most fascinating thinkers and doers, who are challenged to give the talk of their lives (in 18 minutes or less). The best talks and performances from its conferences are made available to the world, for free. More are added each week and all are subtitled in English, and many in other languages, too. These videos can be freely shared and reposted.

TED Talks can be downloaded from http://www.ted.com/talks.

Naked Scientists is a weekly radio programme with a website that offers articles and podcasts that address issues of general interest, such as: "Why do insects buzz around lights?", "Can a mother's diet affect her baby's genes?" and "The Magic of Binary". The Naked Scientists, available online at http://www.thenakedscientists.com, describe themselves as:

> A media-savvy group of physicians and researchers from Cambridge University who use radio, live lectures, and the Internet to strip science down to its bare essentials, and promote it to the general public.

Courses

One-off talks by scientists can be highly stimulating and provide insight into how professionals actually think and talk. The human encounter adds to what is gained from reading. For people who decide to go more deeply into an area of science, courses of study are available at various levels. Adult education services from the local authority may offer some, though the trend has been away from personal interest courses towards qualifications and skills for the workplace. Courses are often set out in a brochure provided by the local authority; local libraries usually have good information about these. Voluntary sector organisations play an important part in local provision for the interested adult. The Workers' Educational Association (WEA) offers courses in topics such as geology, astronomy, chemistry, psychology and evolution. The National Federation of Women's Institutes (WI) runs residential courses[viii] in areas of science such as garden science, forensic psychology and astronomy, in addition to speaker events at its local branches. The University of the Third Age (U3A) provides courses by older people for older people in topics such as botany, astronomy, computing and geology. These are organised locally by volunteers and can be found by searching on the web. Information on some providers is given below.

Courses

Workers' Educational Association

Through curriculum themes of employability, health and wellbeing, community engagement and culture, the WEA gives students the confidence to learn new skills, live healthier lives, engage in society and broaden their horizons. Courses are created and provided through regional offices and volunteer-led branches, often in partnership with local community groups and organisations. Previous knowledge or qualifications are not necessary to join most of courses, only a willingness to share with others your curiosity, ideas and experience.

https://enrolonline.wea.org.uk/online/coursesearch.aspx

(Continued)

(Continued)

The WI's Denman College

The national college of the WI, Denman offers a wide variety of day and residential courses in cookery, craft and lifestyle. Set in the picturesque village of Marcham, Oxfordshire, the Georgian mansion and seventeen acres of grounds provide a good atmosphere for learning.
http://www.denmancollege.org.uk

Field Studies Council

The Field Studies Council (FSC) offers residential and day courses that introduce the natural world through exploring and understanding the special character of the landscape and wildlife around FSC centres. Topics include natural history and walking, geology and scenery.
http://field-studies-council.org/individuals-and-families/natural-history/exploring-the-natural-world.aspx

University of the Third Age

The U3A movement is a unique and exciting organisation that provides, through its local groups, life-enhancing and life-changing opportunities. Retired and semi-retired people come together and learn together, not for qualifications but for its own reward: the sheer joy of discovery.
http://u3a.org.uk

Open University

The Science Faculty offers a series of short courses covering interesting contemporary topics in science such as astronomy, plants, human genetics, nuclear energy, volcanoes, nutrition and the frozen planet.
There are two different types of course:

- **Short modules for credit** are assessed and can be studied either as standalone courses or as part of an Open University qualification.
- **Online short courses** that are more flexible but cannot count towards an Open University qualification.

http://www.open.ac.uk/science/main/studying-science/science-short-courses

Some colleges of further education also offer courses. Increasingly, these are geared to formal qualifications, including GCSEs and A-levels, but in some cases leisure courses may also be available. Some universities offer short courses, evening classes or summer schools in a variety of subjects (see, for example, the University of London short courses[ix]). These can be found through the individual websites of colleges or universities. The Open University in particular offers a range of courses, some of which may be studied for interest and others as part of a qualification.

Festivals and open days

The resurgence of popular interest in science has also led to an increasing range of more active ways to engage with science, through festivals and open days. The British Science Festival[x] moves around the country from year to year, running for a whole week, offering all manner of exhibits, talks and demonstrations. Equally exciting is the Royal Society's summer festival,[xi] held in its prestigious London headquarters in Carlton Terrace Gardens. It also involves scientists in direct contact with the public plus a host of practical demonstrations, often of the latest research. See the next page for further details.

As well as major festivals a number of organisations periodically offer open days. An example is the Institute of Making[xii], part of University College, London, which offers a themed day every month in the area of materials. Hands-on activities have involved materials such as wood, string, foam and plastic. Other science- or technology-based organisations, such as power stations, research institutes and water-management facilities may offer open days, in country-wide locations.

Places to visit

Museums focusing on topics related to science and technology are proving increasingly popular. Some, such as the Science Museum in London and the Museum of Science and Industry in Manchester, are dedicated to the subject. Many others in a variety of fields such as photography, manufacturing, power generation, fisheries, transport and textiles cover important elements of science and technology. The Royal Institution in London, with its two-hundred-year-old mission "for the speedy and general diffusion of the knowledge of

Festivals

British Science Festival

The British Science Festival is one of Europe's largest celebrations of science, engineering and technology, with over 250 events, activities, exhibitions and trips taking place over a week in September. In a different location every year, it offers activities for families and schools groups, adults looking for entertainment and stimulating debate, as well as events aimed at professionals interested in the latest research.
http://www.britishscienceassociation.org/british-science-festival/about-festival

Royal Society Summer Science Exhibition

The Royal Society mounts a science exhibition every summer, which showcases the most exciting cutting-edge science and technology research. Usually based in the magnificent premises of the Royal Society in central London, it provides a unique opportunity for members of the public to interact with scientists and ask them questions about their work. It is open to members of the general public as well as students and teachers, scientists, policymakers and the media.
http://royalsociety.org/summer-science

National Science & Engineering Week

This is a ten-day national programme of science, technology, engineering and maths events and activities across the UK aimed at people of all ages. Programmes vary from year to year and details of activities in each region can be found on the website.
http://www.britishscienceassociation.org/national-science-engineering-week

all new and useful improvements", houses a number of exhibits of importance in the history of science.

In addition to museums, visits can often be arranged to science-based facilities in any local area. Power stations, local manufacturers, hospitals, botanical gardens, wildlife reserves, food and water-quality facilities are obvious examples. University laboratories and research institutes have proved a major resource for the discussion groups. A visit is usually arranged after an issue has been discussed extensively and the need for more specialised knowledge has become clear. A brief search under "research" at a university website provides links to particular experts. Researchers are generally only too willing to receive a visit and to offer a short talk and lab tour. Examples of visits made in this way by discussion groups include cancer labs, an endocrinology clinic, an earth science department, a particle physics team, a musical performance lab and a neuroscience MRI scanning lab. Scientists have been invariably thoughtful about communicating appropriately with non-specialists and about interacting with the questions put and points of view expressed by the visiting groups. The outcome has not only been deeper understanding of the issue in hand but also a special insight into the working life and human characteristics of practising scientists.

Outcomes

Getting into science has been described thus far as a kind of journey that begins with simple curiosity about the natural world and leads on to the pursuit of questions drawn from daily life. Insights gradually develop as a result of reading around, listening to broadcasts and perhaps entering into discussion with groups of like-minded people. Deeper understanding follows through exploring public resources available through the web, museums, talks, festivals and visits. For anyone thinking of starting out on such a journey the question arises: after all this activity, what do people make of the insights and knowledge they have gained?

Gaining confidence

The first point to make, an obvious one, is that this kind of informal approach to learning is not itself a preparation for a science-based career. For a few it may well be an inspiration to move in that direction, but in general it does not have a vocational aim. Its fundamental purpose is a liberal one: to enlighten the individual. For an area of knowledge that was previously dark this can have much wider effects than simply knowing more. Some of these have been identified through focus group discussions with participants in the author's groups. Like any positive learning experience, learning about science "expands your thinking" and "allows you to follow up interests" to quote two participants. Confidence builds up and "it takes away the intimidation aspect of science". As one person put it, "You can see that another person's apparent knowledge can be just superficial." "You can listen to a programme or podcast with confidence."; "If you pick up something about science, now you read it." This boost to confidence leads to further effects: it "opens up doors"; "You have a bit of knowledge about news items."; "You get to talk about new topics — smart materials, genetics etc."

These comments from participants confirm that the outcome is not so much about being able to apply newly acquired knowledge directly in practical situations but it is about bolstering the ability to respond to news, to participate in conversations, to think through issues that have a basis in science. In other words, it helps to put people's acquaintance with scientific concepts on a par with others they have about politics, religion, psychology, geography, social affairs and so on.

Getting involved

For those that wish to, participation can extend beyond conversation, viewing and reading. Active involvement becomes a possibility. For parents, a stronger grasp of scientific ideas can mean greater involvement with their children's school work. Where science might once have been a no-go area for supporting a child's homework, it can become an area in which both learn together. Primary school science, in particular, deals with the fundamentals: the Earth, motion, weight, electricity, life cycles, circulation, plants and animals; and younger children are, of course, bursting with "why" questions. What better way to respond than to learn with them?

Opportunities to participate extend well beyond the home. Societies exist in many localities for particular fields of interest; astronomy is one favourite, geology another and wildlife a third. Many societies are welcoming to new members and different levels of knowledge are usually expected and accommodated. Once self-confidence has begun to build, it is easier to hold one's own in a group of long-standing stargazers or bird watchers.

In recent years the idea of citizen science or crowd science has begun to take hold. Scientists in some fields have realised the potential of large numbers of citizens assisting with massive data collection. The Citizen Science Alliance,[xiii] for instance, organises projects about the Moon, planets, black holes, whales, bats, plankton, cancer, genetics and more. Scientific projects that citizens can engage in at any given time are listed on the web. Some examples at the time of publication of this book are given on the next page.

Citizen Science Projects

Dozens of projects that citizens can engage with are in operation at any given time. Here are a few examples.

Nature's Calendar

Volunteers record the signs of the seasons where they live — the first ladybird or swallow seen in your garden in spring, or the first blackberry in your local wood in autumn. The science of phenology (the study of the times of recurring natural phenomena), especially in relation to climate, involves recording when you heard the first cuckoo or saw the blackthorn blossom and comparing it with other records.
http://www.naturescalendar.org.uk/survey

Verb Corner

Verb Corner is trying to work out what verbs mean. The problem is broken down into a series of tasks, each of which asks about a specific component of meaning that scientists suspect makes up one of the building blocks of verb meaning. This massive project uses citizens and is expected it to make a valuable contribution to linguistics, psychology and computer science.
http://www.scientificamerican.com/citizen-science/the-verbcorner-project

Old Weather

By transcribing ships' logs, weather observations made by US ships since the mid-nineteenth century are recovered. These contribute to climate model projections and improve our knowledge of past environmental conditions. Historians also track past ship movements and tell the stories of the people on board.
http://www.oldweather.org

How Do Galaxies Form?

NASA's Hubble Space Telescope archive provides hundreds of thousands of galaxy images. To understand how these galaxies, and our own, are formed laypeople are helping to classify them according to their shapes — a task at which many brains are better than even the most advanced computer.
https://www.zooniverse.org/project/hubble

An alternative way to get involved with scientific activity is through the growing "public engagement" movement. Universities in the UK are being encouraged to organise activities that enable their research to be shared with the public. The National Coordinating Centre for Public Engagement defines it as a "two-way process, involving interaction and listening, with the goal of generating mutual benefit".[xiv] A range of organisations also run public consultations on scientific issues, particularly ones for which the public interest is strong. Typical topics include safety of food, ageing, genetic modification, brain science and cell therapy. These are, by nature, short-lived, and information about them is best found by searching the web.

Seeing patterns

Beyond the practical consequences of trying to engage more deeply with science, lies another more intellectual outcome. Patterns begin to emerge over time even though concepts are introduced haphazardly as they arise rather than being built up systematically. Perhaps the most significant of these is the realisation that not all concepts are equal — some are more fundamental than others. Thus, for example, underneath a vast range of everyday topics — from refrigerators and electricity supply, to digestion and plant growth — lies the unifying concept of energy and its transformation. Whether in electrical form or as heat or pent-up in fuel, energy is a concept with enormous power to explain everyday phenomena. In a similar way the structure of atomic and molecular structure underlies a huge range of phenomena, from the chemistry of household substances through the behaviour of electric current to the interaction of proteins and DNA in living matter. In the life sciences, understanding of genes and their role in evolution serves a similar fundamental role in explaining reproduction and the evolution of the species.

To give some indication of the range of fundamental concepts that can arise in science discussions, a list is given below of ones that have been explored in discussion groups over the years, each having arisen from an everyday starting point.

- Atomic and molecular structure
- Heat transfer
- Energy

- Forces and motion
- Gravity
- Gases, solids and liquids
- Waves in general
- Electromagnetic waves
- Magnetic and electrical fields
- Electricity generation and flow
- Plate tectonics

- Elements and the periodic table
- Chemical bonds
- Chemical reactions

- Cell structure and contents
- Nervous system
- Hormones and neurotransmitters
- Brain regions and function
- Genetics and DNA
- Evolution

- Human evolution
- Race and culture
- Child development
- Mental health
- Musical theory and performance
- Populations, averages and probability
- Binary arithmetic and digitalisation

It's important to remain patient about the many gaps that any of us inevitably have in our understanding at any given stage. It is paradoxical that the more one learns about science the more is revealed about what you have yet to find out. I suspect this is largely true of any intellectual endeavour and has to be accepted philosophically. People at every level of scientific training are aware of huge gaps in their knowledge; learning is in effect an unending journey, so it may as well be made enjoyable and sustainable, if possible.

Managing without mathematics

Mathematics is widely perceived as essential to understanding science, and many see a lack of mathematical background as a barrier to engaging with science. However, the experience in discussion groups is that much can still be achieved by discussing ideas without recourse to mathematics, though, of course, some areas of science are more open than others.

Some aspects of mathematics do seem to have been internalised by most people in discussion groups. Basic concepts of proportionality seem to be widely understood — the idea that doubling the speed doubles the distance travelled, for example. Graphs and charts are also relatively well understood, perhaps because they crop up in so many branches of professional and everyday life today, as do fractions, percentages and decimals. However, a sense of relative size may need to be brushed up — the fact that a billion is a thousand times bigger than a million, for example.

Some areas of mathematics actually become the subject of interest — for example, binary numbers arise naturally in relation to understanding how digital technology works. What might have been perceived at school as a difficult and abstract area of knowledge can re-emerge as a straightforward and indeed interesting topic. A similar discovery can occur with trigonometry when discussion develops around measurement, perhaps in relation to the extraordinary building constructions and astronomical observations of ancient civilisations. Algebra, however, presents real difficulties for most people: very few feel confident about representing things abstractly through letters and using them in equations. Occasionally a simple equation might be quoted just to demonstrate a dependency ($E = mc^2$, for example) but very few people are in a position to follow the use of equations as part of an explanation.

Different areas of science depend on mathematics for explanation, but to differing degrees. Some topics in, for example, child development, botany or geology can be extensively explored without use of algebra, while others, such as the theory of gravitation or behaviour of gases, less so. Nevertheless, even for topics strongly associated with algebraic explanation, such as thermodynamics or wave theory, some scientists and writers are able to find ways to explain the ideas clearly and usefully without recourse to mathematics. Having said this, deeper and more satisfying levels of explanation do become possible through the use of mathematics, so getting to grips with the subject in later life is always a worthwhile option for the curious adult.

Summary

This chapter has considered some of the opportunities for further exploration once the appetite for science has begun to develop. Of course, the resources made available for adults to engage with science in informal ways are more limited than those provided for young people in full-time education. But nevertheless there are many more than is commonly realised. The obvious ones, such as museums and libraries, are a good starting point, but other resources such as laboratories, wildlife parks, power stations and research institutes can also be looked to for talks, visits or exhibits.

The value of joining together with other like-minded souls is enormous, but it is not common (yet!) to find local discussion groups that focus on science. If you would like to try setting one up, you can look at the suggestions offered on the website that accompanies this book (http://gtgwithscience.com) where you can also contact the author.

Endnotes

i. Open University materials are available online at: http://www.open.edu/openlearn/science-maths-technology/science.

ii. Carlton, S. (2001). *Life, The Universe and Almost Everything: The Value of Adults Learning Science.* Leicester: NIACE. Available at: http://shop.niace.org.uk/life-universe.html.

iii. For Institute of Physics public events see: http://www.iop.org/events/public/index.html.

iv. For Royal Society of Chemistry events see: http://www.rsc.org/conferencesandevents.

v. For Society of Biology events see: https://www.societyofbiology.org/events.

vi. *Little Atoms* podcasts can be downloaded from: http://feeds2.feedburner.com/littleatomspodcast.

vii. *TED talks* can be downloaded from: http://www.ted.com/talks.

viii. For Women's Institute residential courses see: http://www.denmancollege.org.uk/courses.

ix. For the University of London short-course programme see: http://www.london.ac.uk/2114.html.

   x. See: http://www.britishscienceassociation.org/british-science-festival/about-festival.

  xi. For Royal Society summer festival see: http://royalsociety.org/summer-science.

 xii. For Institute of Making see: http://www.instituteofmaking.org.uk/events/series/public-events.

xiii. For Citizen Science Alliance see: http://www.citizensciencealliance.org/index.html.

xiv. For the National Coordinating Centre for Public Engagement see: http://www.publicengagement.ac.uk/what.

Chapter 9

About Science Itself

Introduction

The main purpose of the approach described in this book is to help people get to grips with important scientific ideas. The focus is on aspects of science that relate to the questions people pose. However, a secondary strand also develops out of such discussions, in parallel with the substantive topic. People become interested in the nature of science itself — its methods, its history, the culture and ethics that surround it. Discussions of this kind prove particularly interesting not only because they cover important themes but also because they enable participants to draw upon their own life experience and professional knowledge.

With hindsight, the emergence of this secondary interest is not so surprising. Disengagement from science in the teenage years results not only in poor grasp of scientific concepts but also in a lack of understanding of the nature of the subject itself. The basic tools to help navigate the subject — its disciplines, methods and history — may simply be missing. Yet, paradoxically, the issues of scientific method, historical development of theory and the practical use of evidence seem to be of particular interest to people in discussion groups.

This chapter addresses a number of these aspects of science as they have emerged over the course of some 200 free-flowing discussions. It does not purport to cover all aspects of the nature of science, but rather to illustrate a range of issues that have proved important to groups of laypeople. The headings that follow reflect the types of issue that arose regularly in discussion.

History

The emergence of our modern concept of science from earlier forms of philosophy and religion is a popular topic, drawing on people's recollections of

TV documentaries, books and sometimes subjects people may have studied at some point — history, philosophy, religion and literature, for example. The way in which concepts have developed historically is not only interesting in itself, it can also help to clarify ideas and misunderstandings people have as they learn about science today. An example of an actual dialogue of this kind is given on the next page.

In a related discussion the following question arose: "What caused chemistry and physics to split as subjects when previously there was just natural philosophy?" This led to discussion of the development of the modern subject of chemistry from the original Arabic science of alchemy and to the role of observation and meticulous measurement of the weight of substances in the late seventeenth and eighteenth centuries. The development of accurate weighing machines was introduced as one of the factors in the growth of chemistry during the Enlightenment period. As a member of the group pointed out, it is often technology that leads the way to advances in science. In a different discussion about Galileo's contribution to the Sun-centred model of our planetary system it was pointed out how vital developments in lens grinding had been in enabling a telescope to be made capable of observing detail in the heavens. A more contemporary example followed a visit to a brain science laboratory; the development of the MRI scanner, which reveals the location of areas of activity in the brain, was described as central to recent advances in neuroscience. The role of technology is a recurring theme in historical discussions and extends to the economics and politics of big science today. "What is the purpose of the Large Hadron Collider?" was once the opening question in an exchange of views about the ultimate quest for knowledge.

Galileo's dispute with the church authorities in Rome gave rise to a discussion, one of many, about the conflict between new science and prevailing ideas. Brecht's play about Galileo was recalled by a member of the group who made the point that before the growth in experimentation, religion was in effect the science of the day, the system for explaining the natural world. The play's point that the Vatican thought the peasants' belief in a better life hereafter would be undermined by evidence that the Earth was not at the centre of the universe, led to exchanges about comparable issues today. Is it ever right for people to go on with wrong understanding, like creationism? How do religious scientists reconcile their belief in evolution and scripture? Isn't

Historical Development of Science

The importance of "wrong" theories

An idea in ancient Greek philosophy, in the area we now call physics, was outlined by the tutor. In ancient Greece it was believed that objects on Earth tended to fall to Earth and come to rest, whereas things in the heavens — the planets, Sun and Moon, for example — continued to move unaided. It was thought to be in the nature of objects to do so; the heavens are perfect, Earth is flawed. One member of the group commented that people must have known that things that fall further will hit you harder. Another suggested there may have been no concept of acceleration — things that fall from higher up might just give you a bigger bang when they hit you!

It was pointed out by one participant that there was no tradition of measuring things or doing experiments at the time. The prevailing approach was to reason things out; so, for example, the heavens are perfect, spheres are perfect, so heavenly bodies must lie on spheres. The social dimension was introduced by another contributor: work was for slaves and experiments were like work in ancient Greece; civilised people sat around and thought.

Another member of the group who had read around the period said that Ptolemy (an Alexandrian) had made measurements. Ptolemy had observed the heavenly bodies and saw that they didn't fit the theory, so invented the idea of equants — points inside the Earth away from its centre, about which the heavenly bodies rotated. This device made the theory fit the observations better but now we know it was simply forcing a wrong theory (the idea that the planets revolved around the Earth) to work better. At the time it was a brilliant advance and it did improve theory; it also introduced a little bit of imperfection.

This point moved the discussion on to the role of "wrong" theory; sometimes it can still be very important in moving understanding forward. This was demonstrated when a participant remembered that Copernicus was a crucial intermediate figure who got away from the "Earth at the centre" idea even though he wrongly believed that planetary orbits must be perfect circles.

One person had watched a TV documentary about Islamic science which had shown how important the Islamic contribution had been in gathering together the documents from ancient Greece, looking after them, studying them and transmitting them to the West. Even though much of what they said was technically "wrong" in modern terms, the actions helped keep alive the idea of scholarship and reasoning.

religion important in holding our mental health together, as the evidence on factors affecting wellbeing suggests?

Discussion of "wrong" theories has led on many occasions to consideration of how understanding develops over time. The book *The Sane Society* by Erich Fromm led to a discussion about distinguishing those ideas of Sigmund Freud that are still seen as valid today from those that no longer are. One member of a group saw some key Freudian ideas, such as a subconscious built from repressed feelings, as still important even though many, such as the Oedipal theory, no longer stand up cross-culturally.

Another conversation focused on theories that proved to be ahead of their time rather than wrong. It emerged unexpectedly in a discussion about computer software in which the use of binary numbers to code information digitally was being explored. On discovering that binary arithmetic had been discovered as long ago as the seventeenth century one participant exclaimed, "It's extraordinary how abstract ideas like binary numbers turn out to be vital many decades or centuries later. It's the same for atoms: they were an abstraction invented by the ancient Greeks and only taken as real in the early twentieth century."

The way in which a new theory replaces a previous one was the focus of one discussion on the historical development of theory. It is summarised on the next page.

The role of the individual in relation to the collective accumulation of knowledge over time has been the subject of many discussions. The discussion summarised on page 107 in Chapter 6 recalls how Mendeleev's breakthrough in creating the periodic table had depended on the work of dozens or perhaps hundreds of scientists over the centuries making meticulous measurements of countless chemical processes. He observed patterns in the nature and behaviour of different elements which ultimately enabled him to represent them in a single tabular form. A great individual had used his imagination plus the accumulated data of many others to make a momentous contribution to science. This reminded one member of the discussion group of the Great Man theory of history; as she put it, "Did Abraham Lincoln personally alter the course of US history?" Was Mendeleev a kind of Lincoln, or was he just lucky to be around at the right moment in history? Was the periodic table simply waiting to be discovered at that moment?

Historical Development of Theory

Radicals vs the establishment

A member of the group had seen "a brilliant documentary" that traced the development of atomic theory and quantum theory. It set out the struggle between the radical physicists Bohr and Heisenberg on the one hand, who claimed that quantum theory was revolutionising everything we think about the nature of reality, and on the other hand the more established pair, Einstein and Schrodinger, who thought it ran against intuition. This led to questions about innovation:

> "Why do people fear new ideas?"
> "Does the mind have to close as you get older?"
> "How did Bohr get his outlandish idea of quantum leaps — did he simply pick up a whacky idea from Max Planck who was working in a completely different field (heat radiation)?"

People pointed out that the struggle between radicals and established people occurs in many spheres of life. One person had read about a distinguished early twentieth-century professor of geology who had feared that if the theory of plate tectonics turned out to be valid, then seventy years of his work would have been wasted. It did! The personal consequences of scientific development can be tough.

The role of intuition was raised: "It's undervalued; people don't trust it, don't like Eureka moments," said one participant. "Unusual connections between ideas get made at strange random times — down time, like the moment of waking up," said another. The tutor pointed out that scientific work is often pursued along the "wrong" path but, of course, nobody knows in advance which path is going to turn out to be right. This can hardly be described as waste — it is the way of science. Robust knowledge builds up slowly from myriad tiny contributions, many of which turn out to be false trails.

People in the group commented on the similarity with other fields, citing as examples artists like Monet, who were revolutionaries and were originally rejected by the art establishment. "Scientists and artists are both challenged by society — maybe because they scare people with unfamiliar things," one person put it. But, as another added: "Science is distinctive in that it sets out propositions that can be tested."

Questions of this kind have led on occasion into discussion of Thomas Kuhn's ideas, set out in his groundbreaking book, *The Structure of Scientific Revolutions*.[i] By studying the way big new scientific ideas emerge, Kuhn showed that scientists tend to work normally within a prevailing paradigm of accepted knowledge most of the time; it is only rarely that a revolutionary moment occurs in which the paradigm is changed. After such a moment, established scientists tend to try to refute the change and sometimes disbelieve it to their death, whilst younger scientists (or those new to the field) gradually take it on and begin to operate routinely in the new paradigm. Truly major advances happen incredibly rarely and very few scientists have the privilege of being involved in one. Kuhn's thoughts bring home to non-specialists the reality of science as it is actually practised. They help to demystify science by bringing to the fore the human aspects of scientific advances, characteristics that resonate with people's broader experiences in life.

Uncertainty

One consequence of learning about science as it developed historically is realising how uncertain it is at any given time. As a member of a group once said:

> Science comes across as too certain. At school it puts across certainty then later says it is wrong. In other subjects, like history, you also learn in different ways as you mature; maybe science is just too certain in the first instance.

A colleague added: "Certainty is understandable in maths. Maybe the paradigm in science is wrong — it puts itself forward as the truth." In another discussion one person talked about one of her teachers who had "got more buy-in from students when she invited them to find out why an experiment hadn't worked". Others remembered being told: "If you don't get the right result from an experiment do it again till you do." One participant with a background in sociology felt that:

> Science represents itself as solid when it is really fluid. Society and culture and meaning are always there. In nineteenth-century Vienna, Freud had to represent his work in positivistic terms because that was the prevailing ethos in his day. Perhaps he invoked Oedipus because that myth was acceptable at the time.

Getting directly in touch with contemporary scientific research is perhaps the most convincing way to demonstrate the inherent uncertainties in science. A visit to an astrophysicist at a local university demonstrated this vividly when she explained that the concept of dark matter had been introduced to explain why rotating galaxies didn't seem to be flying apart — there simply had to be much more matter in the universe than could be accounted for. One member of the visiting group asked quizzically: "So dark matter is just a supposition then?" "Yes," came the reply, "and some people believe we may have to revise the theory of gravity to account for it. There is some incompatibility between it and quantum theory." Every visit, whether to a neuroscience lab scanning teenage brains, a geology department crushing deep rock samples or an endocrinology clinic researching appetite hormones, plunges the visitors straight into the boundless realms of the unknown and uncertain.

It's not only theories that can appear less certain than one might expect, even definitions can change over time. In one discussion triggered off by an article in *Scientific American* it transpired that the very basis on which species are defined has recently changed from descriptors such as "having a backbone" to the degree of similarity in the DNA sequence. In similar fashion, a group member reported the news that the definition of a planet has been changed as a result of decisions made at an international conference. Apparently Pluto is now a ball of dust or ice rather than a planet. Once again the image of science as rigid and unchanging is made more realistic through engaging with actual science as it happens. As one person expressed it in a discussion:

> We have got to consider how the ways in which things are defined scientifically may themselves change. This can come as quite a shock if you believe science is about immutable facts.

In an interesting contrast, one rather profound discussion explored areas where absolute certainty is indeed assumed. Conversation started with an input by a member of the group who had tried reading *Why Does E = mc²* by Brian Cox and Jeff Forshaw. Discussion developed through a series of fundamental questions about atoms and elements:

> "Do all carbon atoms have six protons?"
> "Could we get a surprise by finding new elements on another planet?"
> "Will different planets have different nuclear forces — after all they have different gravity?"

The outcome was a realisation that, in contrast to the uncertainties of experimental science, some theories are understood to be universally valid at all times, in all places: all carbon atoms do indeed have six protons, wherever they are in the universe; gravitation acts identically everywhere; the speed of light in a vacuum is everywhere the same.

Uncertainty not only features in discussion about the practice and history of science, it also features in discussions about the politics of science. An example of this is given below.

Dealing with Complexity

Technical and social factors

On a trip to Alaska a member of one discussion group had seen glaciers that are collapsing and found that some experts describe this as natural, a feature of the receding ice age we are in, whereas others count this as evidence of global warming due to human activity. On returning, she asked the group, in exasperation: "Why is science so controversial?"

Inherent difficulties within science itself provided the first level of explanation. Phenomena like climate change are the result of many simultaneous factors — the changes in the composition of the atmosphere, the behaviour of the ocean, levels of solar radiation and so on. Analysis is highly complex. There is also imprecision in measurements of any kind and theories or models are rarely definitive — there are often several alternative versions.

Discussion moved on to the history of notorious predictions, made in the name of science, which turned out to be false. One participant pointed to the early twentieth-century eugenics movement as an example that had predicted a looming crisis resulting from what some believed at the time to be a lower birth rate amongst the more intelligent. He suggested that the modelling for this prediction could be seen with hindsight to have been flawed — missing out, for example, the effect of increasing female education.

The problem of dealing with complex situations led on to comparing the roles of the natural and social sciences. One person suggested, "Perhaps a scientific approach is where we try to programme in *all* the factors; social sciences operate where we can't". But, as environmental science shows us clearly, even in science we do not know all the factors — yet they still continue to operate.

In summary, science does not have perfect answers; it is principally the art of measuring and modelling. Interpreting the results often involves significant uncertainty. Perhaps the social and natural sciences are not as far apart as commonly thought.

The Experimental Approach

The way in which scientists go about measuring and modelling is a regular feature of discussions about science itself. What is "scientific method"? How are discoveries made? What is it that is distinctive about the natural sciences compared to the social sciences and humanities?

An interesting discussion about this once began with the question: "Could poltergeists exist, but just not be understandable by us? Could psychics be as right as Einstein?" A member of the group pointed out that Einstein had theorised first, using previous knowledge, mathematics and thought experiments; then, crucially, he had used this to make predictions that could be checked. Subsequent observations confirmed the predictions and so validated the theory. Another member of the group extended the idea of scientific method to include studies of human behaviour. She pointed out that in areas such as health care and employment services, experiments are sometimes carried out using control groups, which compare the effects on people receiving some kind of intervention with others who don't.

The singular phrase "scientific method" perhaps gives a misleading impression — in practice, many different kinds of method are used and the choice depends on the context. In chemistry, for example, laboratory experiments are designed to minimise the number of variables acting in a situation so that conclusions can be drawn about cause and effect. In subjects such as astronomy, geology and archaeology, research depends more on exploration, observation and theoretical deduction than experiment. In areas of social science, where the number of variables cannot simply be limited by design, interventions may be tested by assigning participants randomly either to a group that receives the intervention or to a control group that does not. This means that any initial differences between the groups result only from chance so that with large enough groups any differences found between the two afterwards will be due to the intervention. Varieties of this method are also used in some branches of natural science, including agriculture, medicine and psychology.

Interestingly, some of the methods and technologies traditionally associated with the natural sciences are proving useful more widely. Historical dating, for example, depends increasingly on radiocarbon dating using methods from nuclear physics; obstetrics uses ultrasound methods developed in acoustics; even literary analysis now makes use of computer software to measure the frequency of word usage. Science is not simply a subject or a profession; it is the way we approach phenomena that are susceptible to measurement and experiment in any field.

If observation and experiment are the means of testing a hypothesis, how do hypotheses themselves originate? This is an interesting and frequently occurring issue for discussion. Many participants refer to folk wisdom as an important source, e.g. the use of citrus fruit to protect against scurvy or the adoption of traditional medicinal plants as a starting point for developing drugs. However, in reality much of the inspiration for a new experiment comes simply as the next logical step from earlier work. Nevertheless, the unpredictable, even playful nature of some basic science struck one group during a visit to a cell biology lab at a local university. "It seems so arbitrary, messing about with molecules to see if things might work. It's quite shocking. After all, doctors have to be quite precise," was how one participant saw it. In another discussion people commented on the accidental discovery of drugs. One person told the group, "It's amazing — I read that Viagra started out as a heart drug, but when it was trialled some of the males began reporting other, quite unexpected effects!"

Lab visits help reveal, in the words of one visitor, that:

> Science is not so cut and dried or clear about its direction and purpose as we may think. The stereotype is misleading — some aspects of scientific work are very methodical, others are open-ended and require imagination and creativity. And sometimes discoveries are simply accidental.

However, the aim of all sciences, natural, social and human, is to test hypotheses as rigorously as possible. This is as true for an analysis of the causes of the First World War as it is for an experiment on the role of hormones in anorexia. Ideally, the intention of a scientific experiment is to try to disprove a hypothesis — in this sense, science is sceptical. Successful theories are ones that have survived all attempts at disproof, to date.

A further revelation when groups discuss a piece of contemporary research or visit a laboratory is the way in which different disciplines, with their various methods, are needed. The nature of the problem requires specialists from different fields to work together. In a discussion about the migration of peoples, for example, it became clear that research on the topic involves biochemists to analyse the DNA of individuals, geographers who know about topography and urbanisation, botanists with their expertise in crop growth, and computer specialists who analyse family and place names, let alone the anthropologists whose main field of study it is. A different discussion about continental drift showed how the development of the theory of tectonic plates had depended on the work of oceanographers measuring magnetic fields, geologists drilling out core samples in the Arctic ice and physicists studying the flow of molten rock — each of whom was reliant on the specialist expertise of the others.

Methods that engage the natural sciences with the complexity of human and social affairs offer interesting insights into scientific methods. Particularly fruitful discussions have developed about studies that use identical twins brought up separately. Because of their power to separate out effects due to the genes the twins share from the environmental factors they don't share, such studies throw light on the basis of characteristics such as cognitive ability and sexual orientation. Other scientific methods used in social investigation involve the longitudinal tracing of the lives of individuals. These reveal links between early factors in life, such as parents' education and household income, and outcomes later in life, in areas such as health and employment. Discussions of this kind, about the role of scientific methods in social matters, help to broaden perceptions of science beyond the stereotype of white coats in a laboratory.

Theory Building

Historically, theoretical ideas tend to emerge gradually (though occasionally they explode forth suddenly) from the build-up of insights from measurements, observations, experimentation and imagination. The way in which this actually happens is a popular topic of discussion. The development of theories over historical time provides a useful insight that counters the idea of rigid authority so often associated with the natural sciences. It becomes clear that the concept of "laws of nature", appropriate perhaps in some areas, such as thermodynamics and gravitation, is too inflexible for many others. The concept of models — images, metaphors and equations that explain the way systems behave — is often more helpful in capturing the provisional nature of much knowledge in science. A good model accounts well for known behaviour and can also be used to predict or explain new behaviour. An example occurred in a discussion about the Earth's magnetism. The model of a north–south magnet, based on the flow of molten iron in the Earth's core, was introduced to describe the magnetic effects observed at the surface. It usefully explains the behaviour of a compass needle and aids navigation. But in addition, when streams of charged particles were later discovered emanating from the sun, it explained why they are deflected on reaching the Earth, causing the colourful aurora borealis. A subsequent visit to a geophysics laboratory showed how models often develop over time. Samples taken from ancient seabed rocks had revealed that the Earth's magnetic field had surprisingly reversed direction many times. As a consequence, a more sophisticated model had to be developed to explain this.

The use of models in place of ultimate truths also helps us to understand ambiguity. For example, the nature of light and other forms of electromagnetic radiation appears in some experiments to be wave-like, and in others, particle-like. Similarly, the electrons in an atom are treated in some situations as particles orbiting the nucleus, in other situations as smeared-out clouds of negative charge, and in yet other situations as abstract entities with fixed levels of energy. At school the way in which these alternative models replace one another in successive years can be disturbing for the learner; paradoxically, for adults this very ambiguity may be welcomed because it parallels the way complex ideas are handled in other areas — literature, history and art, for example. Deeper discussion about the nature of reality itself can easily flow from consideration of models — are they all that science can be expected to offer or does a deeper objective reality lie beneath them?

A related insight into the nature of science emerges from discussion of models — the role of idealisation. Idealised situations are often used to develop scientific theory: surfaces are treated as perfectly smooth, air as offering no resistance, stars as point objects. Discussion of models, such as an ideal gas, that enable useful approximations to be made to real events show the value of models that simplify complex realities, yet provide useful predictions of behaviour. A joke told by a member of a discussion group captured the idea nicely:

> How can you predict the outcome of a horse race?
> *Statistician*: Give me all the data and I will work it out for you.
> *Physicist*: We have already worked it out ... for a spherical horse in a vacuum.

Such unrealistic conditions can seem absurd when presented baldly in a mathematically based argument; but, when explored with their merits and limitations in discussion groups, can be appreciated as important for developing theory in complex situations.

A further insight into scientific theory comes from exploring the way theories develop over time. One discussion about genetics revealed that the concept of some factor that enables information about the body to transfer from one generation to the next was theorised well before genes were actually identified. It took many decades of further experimentation to discover that genes were associated with DNA rather than proteins or some other substance. The discovery of the double helical structure of the DNA molecule took several more. As we saw earlier, sometimes what we now consider as "wrong" theories may have nevertheless been useful in their time. As a member of a group once said:

> Copernicus believed wrongly that planetary orbits must be perfect circles, but he was a crucial intermediate figure who got away from the "Earth at the centre" idea, even though he didn't get as far as seeing that orbits were elliptical. Kepler achieved this later on.

Positive effects of placebos were once dismissed as a demonstration of the gullibility of patients. But now the placebo effect is taken seriously as an aspect of treatment. Of course, wrong theories can also be dangerous — bloodletting, for example, based on the ancient theory of the four humours, persisted well into the nineteenth century.

The role of imagination in developing theory is often seized upon gladly in discussions, as an indication that science is not purely cool and logical. Images and metaphors seem to help us bridge from what we know to something new. This applies very much as we learn about science and it is always gratifying to discover it holds true for great scientists as well as oneself.

Imagination helped inspire the invention of the periodic table, the theory of special relativity and the idea of universal gravity. Metaphors enable us to describe one phenomenon in terms of another, to use familiar images to explain something unknown. Thus electricity is explained as a kind of flow, like a river, and the word current is adopted; light is described in terms of waves like ripples on a pond, or particles are described as shot from a shotgun. After a visit to a particle physics lab, one group were struck by a researcher who talked about "seeing" a particle when in fact he had simply noted a blip in a series of numbers on a computer screen. They had witnessed the kind of routine metaphorical talk, typical of scientific communities.

Interestingly, discussions can also provide insight into the limits of metaphor. From time to time simple but profound questions are asked that would challenge a scientist accustomed to purely mathematical answers, e.g. "What exactly are gravity and mass?" In responding to questions of this kind it becomes clear that some concepts in science are truly fundamental; they simply cannot be explained metaphorically in terms of other more familiar things — they are things unto themselves. Questions of this type can easily lead to discussion about the inherent limitations of scientific, as opposed to philosophical or religious, explanation.

Using Evidence

Thus far we have looked at ways in which scientific methods and theories have developed over time; but discussion sometimes goes further than this, touching on how evidence from scientific investigations gets used in the practical world. Within the scientific community itself overarching scientific concepts are built up gradually from many separate sources of data collected in specific situations. For example, the science of climate change depends on many quite separate endeavours, often quite uncoordinated — measurements of the levels and salinity of the oceans, of atmospheric temperatures and humidity, of glaciation and ice samples, of wildlife numbers, growth of vegetation and much more. Synthesising evidence of varying quality from diverse contexts is itself a major scientific task. With rare exceptions, scientists are reluctant to make judgements from single, unreplicated studies or to draw out implications from evidence that has not been rigorously synthesised. As a consequence, it takes a long time for sufficient evidence to accumulate to enable definitive conclusions to be drawn. This time lapse can be very frustrating for issues of public concern as the discussion below illustrates.

It can come as something of a surprise to learn how evidence is sometimes put to practical use long before the underlying theory is fully understood. A discussion about the way in which medicinal drugs are developed focused on the way in which evidence is often based simply on the observed effects on animals or humans taking the drug, once it has been tested for toxicity. The precise mechanism by which drugs actually work may be unknown at the time.

Discussion sometimes focuses on the way even good, solid evidence is not always accepted by people and put to use. The historic study of cholera by the physician John Snow in the 1850s was described by one group member. Snow's plotting of the geographical incidence of cholera in London's Soho district

Waiting for Evidence

The health risk of mobile phones

A discussion once began when a member of the group said she was worried that her ear appeared to redden whenever she used a mobile phone. She wanted to know what science has to say about the health risks of these devices; some newspapers were reporting alarming studies. In discussion it was pointed out that it would take time for the evidence on the health effects of electromagnetic radiation from mobile phones to become persuasive; for experiments to be repeated in different places at different times and extraneous factors to be discounted. At any given time the jury is often out on key health questions. But different interest groups may pounce on single studies that favour their preferred view and thus neglect contrary ones; hence the need for synthesis of evidence from multiple studies. But others pointed out that some of the evidence about climate change began to appear twenty to thirty years ago yet was not heeded. Debate about the denial of scientific evidence led on to scientific ethics — the responsibilities of the media in reacting to single studies and the role of the scientific community in trying to replicate results and peer review studies.

eventually led to the removal of the water pump at the centre of the outbreak and a drop in the incidence of the disease. Despite this the authorities were apparently not persuaded to accept the radical germ theory of disease and take appropriate preventative action. However, as others suggested in further discussions, failure to act on evidence is not such a remote misdemeanour. In our millions, we eat foods known to be unhealthy, take minimal exercise and in some cases drink and smoke in defiance of scientific evidence. Psychologists and neuroscientists are today studying the all-too-human reasons for this reluctance to act rationally on the basis of evidence.

The use of scientific evidence in policy is a regular topic of discussion, presumably because it features prominently in media reports whenever it arouses controversy. Discussion groups include people from many walks of life, so insights into this political aspect of science can come from a variety of professional perspectives, as the example below demonstrates.

This example echoes the point made by the government scientist described in Chapter 5, that scientific research informs rather than dictates when other factors such political judgements have to be taken into account. In his view better education in statistics for all would go a long way towards improving public policy.

Use of Evidence in Policy

Classification of drugs

The case of a scientific adviser on recreational drugs had been in the news; he had recently been sacked by the secretary of state after his pronouncements on the relative risks of using cannabis. A member of the discussion group who had knowledge of health and safety policy was aware of the mathematical basis of risk: she explained that people do not always respond rationally to levels of risk. For example, cancer deaths far exceed car deaths, yet smoking is not outlawed. But government did insist on bigger health warnings on packs so policies can mitigate dangers. She pointed out that although science is expected to provide the best possible factual evidence on political issues, the determination of policy has to take other factors into account as well, such as public perceptions, media reactions and costs.

The use and misuse of statistics is a recurrent issue in discussion groups. In a discussion with a visiting statistician about the use of league tables in hospital closures it was pointed out that the error intervals can be so wide that cases in the middle region cannot truly be distinguished. This can lead to unreasonable decisions based on league table position, in which the large margin of error is ignored.

The distinction between individuals and populations is another recurring statistical discussion point. In one example, discussion centred on research in neuroscience that showed a link between educational attainment and a particular aspect of brain activity that differed in children according to socio-economic background. In discussion, one member of the group said:

> Physical explanations of psychological things are mediated socially and people may apply what they hear about average values for a whole population to the individuals they come across. Even some fully trained scientific experts may simply not be aware of the social class implications of what they are exploring.

A further aspect of statistical interpretation occurred in a related discussion about research design. It was pointed out that:

> In any experiment or study someone at some point has to choose what variable to look at, to see if it relates to another one; for example, the researcher chooses whether to look at social class or income or whatever. What about all the other factors you didn't choose to look at, like the role of breastfeeding, for example?

The ensuing discussion focused on the way media reporting can exploit this. A story may focus on a finding that, say, Yorkshire has the greatest number of obese people in the UK (a purely fictional statistic!); yet this may simply reflect the fact that geographical region was chosen as a variable to report when other variables, such as age or genetic make-up, may have had far greater relevance to health.

The application of scientific evidence in policymaking and the way in which statistical information is used are particularly valuable areas for discussion. They engage people actively because they are often grounded in contemporary news stories and also enable important statistical ideas to be explored carefully.

Ethics

The way in which scientists interact with the wider world of media, policy-making and commercial interests introduces another direction that discussions about the nature of science sometimes take: the issue of ethics. Occasions on which scientists appear to have acted unethically feature prominently in the news; this provides an opportunity to discuss the behaviour of scientists and scientific culture more broadly. An example of such a discussion is given below. It began with a topical event — a volcanic eruption — and then moved on to the broader issue of climate change.

Ethical Behaviour

Emails about climate science

The press was awash with stories about emails from a British university that had been interpreted by some as an attempt to suppress contrary evidence. The first point made in a discussion of this suggested that scientists, like others, are sometimes prone to exaggeration; as one person put it: "Science is human like other endeavours, with all human weaknesses." Another recalled reading *The Double Helix*, James Watson's story of the discovery of the structure of DNA: "It shows the human side of science — competitiveness and deceit — but, distinctively, scientific claims are subjected to scrutiny, through experiments." A third member of the group recalled reading that Darwin himself had "been up against huge resistance in developing his theory of evolution". A fourth, a former journalist, added that "today the media make it even tougher, but journalists will never miss a chance of a scoop".

Members of the group recalled various cases of scientists alleged to have manipulated data — one mentioned Sir Cyril Burt and data on IQ, another referred to Robert A. Millikan, whom is said to have withheld some data in his influential "oil drop" experiment to measure the charge on an electron. Others referred to unethical activity such as the publication of fraudulent research on the MMR vaccine that led to a rise in measles and mumps.

Discussion moved on to what is to be done if even scientists have flaws. One person pointed out that "football has fouls, but there are referees who adjudicate". The peer-review process was discussed with its procedure for ensuring that the work of an expert is scrutinised by other experts.

Because incidents of unethical behaviour are much more widely reported than normal scientific activities, public perceptions are inevitably coloured by such events, rare though they may be. Discussions about the nature of normal science and visits to practising scientists help to rectify this bias. An illustration of this occurred when a group of scientists in Italy reported a bizarre finding that a particle appeared to be travelling faster than the speed of light — a strong challenge to prevailing theory. The scientists had tried all means they could to disprove the finding without success, so went public, to invite other scientists to look for an error in their procedure. This sceptical attitude, in which the scientific community makes exhaustive efforts to *disprove* a hypothesis, is encountered as matter of course on visits to scientists. In a visit to a neuroscience lab the neuroscientist was asked questions by the group that are relevant to everyday life but very broad in nature. She was careful to answer only in relation to specific research findings where the evidence was quite certain. For example, when asked by a mother, "Why are adolescent boys so difficult?" she replied:

> The social effects of puberty differ in different societies. Thirteen-year-old boys seem less able to read some signals about behaviour. Some synapses that have grown in childhood but are not used seem to get weeded out at puberty.

She wisely countered the questioner's desire for an all-embracing answer with the cautious response that "we don't know the whole story, but we do have bits of evidence from specific studies".

Scientists are not only under pressure from the public to provide simple answers on complex issues, but may also be subject to commercial pressures. One discussion focused on the history of patronage — who actually pays the scientist? In Galileo's time, the prince was the key patron for scientists as much as for musicians and artists; indeed, it was the Grand Duke of Tuscany who appointed him professor at the University of Pisa. The military have also been important patrons of science throughout history. A member of a group who had been reading about Galileo pointed out that at one stage in his life he had been supported by the Venetian Arsenal because of his contribution to the science of navigation. Discussions have also brought out the importance of manufacturing industry and economic development in the development of science. Typical discussions have focused on the growth of steam power in the early nineteenth century transforming our understanding of thermodynamics, and the laying of telegraph cables under the Atlantic boosting the study of electromagnetism.

The contribution of manufacturing and commerce to scientific progress also stimulates discussion about the influence of commercial interest on scientists today. Discussion with researchers on visits to laboratories has thrown up some of the complexities of this. On one visit, to a lab working on how biological cells signal to one another, the senior scientist was asked about patents. He told the group that patents do impose restrictions on research but are needed to encourage companies to risk investments for the next phase of research and development (R&D). The delicate balance was illustrated again on a visit to another cell biology lab where the researcher was asked whether scientists tended to share their data. He described two different views:

> Either you don't share till your results have been fully tested, or you share them as soon as possible. Either way you do worry about the competition — there are issues of patenting, which may encourage you to keep quiet.

Interestingly, the researcher went on to clarify the important difference between basic and applied research: his own was basic, involving multi-disciplinary teams of geneticists, chemists, physicists and zoologists exploring chemical pathways in the brain and as such data was shared freely. But his discoveries would eventually be passed on to more applied scientists — pharmacologists and medicinal chemists — who might be able to use them to develop potentially useful anti-depressant treatments; their work could well be patentable.

Discussions about commercial interests extend beyond the issue of patents. Ethical issues in food labelling, for example, have emerged from discussions about the effects of fats and sugars on the body. The issue of pricing HIV drugs for the developing world has arisen from discussions about the methods used, and investment required, to develop modern pharmaceuticals. The complexities were illustrated in one exchange about scientifically misleading claims about so-called superfoods: "Can't there be a standard system for alerting customers to the content of foods?" asked one person. "If there were to be thorough legislation on the scientific claims of different dietary products, a huge industry and its workforce would be put out of business," replied another.

Summary

Discussions about the nature of science range over many topics, as this chapter has shown — from methods to ethics, theory to policymaking. A discussion held in a group one day that began simply with a comment about a radio programme went on to cover so many of the issues raised in this chapter that it serves as a useful summary.

A member of the group had listened to a good radio programme on the formation and history of the Royal Society, which explained how it had been created to verify and share information. In the early period, science had been for the upper class — people who had leisure and the money to create a lab and buy equipment. Apparently, Thomas Huxley had been the first "professional" scientist. Later, a different organisation, the Royal Institution, had been created for the benefit of ordinary people. Michael Faraday came from an ordinary background and worked as a technician there. He learned his trade in a practical way and made crucial discoveries in the new area of electricity.

One member of the group mentioned "a huge ruckus at the Royal Society about the teaching of evolution that had led to the departure of the chief education person, who was both a professor of science education and an ordained priest". Apparently, the latter had said creationism should be discussed in science lessons if pupils raise the issue. His experience had led him to believe it was more effective to include discussion about creationism alongside scientific theories such as the Big Bang and evolution, rather than simply giving the impression that such children were wrong.[ii] Another participant commented that the action by the Royal Society shows just how important the political and social aspects of science are.

Another person extended this discussion of scientific culture by raising the question of peer review, also in the press at the same time because of doubts arising about research into climate change that had not been peer-reviewed. Others felt this was a diversion, squabbling over relatively small matters, given the strong evidence about the major threat of climate change. Others expressed wariness about "any blanket claims — the world can be turned upside down by a Newton or Einstein — never say never". "But don't the people against climate change have an agenda?" asked another. "Maybe, but models of climate should be open to peer review. It's worrying that weather stations monitoring temperature in China may have been moved between cooler rural environments and warmer urban environments," replied a third.

Discussion moved on to a general realisation that "it is no longer possible to just say 'hallelujah' — this has been scientifically proven". One person mentioned:

> Experiments in psychology that show that you get different results in trials depending on whether the professionals wear white coats or not. The observer has an impact … Well-meaning pressure groups can create problems — criticising technical solutions and suggesting that only radical changes in lifestyle can make a difference. It is like Puritanism; it can just create guilt in a modern way.

Others felt impatient about the arguing over climate change: "We need both technical solutions and changes in behaviour. Recycling has shown that the latter is possible."

As the debate drew to a close, concern was expressed about the danger of destabilising the whole scientific structure because of the public response to a very small fraction of errors and human foibles. Unreasonable responses to scientific evidence are just one consequence of the damage that has been caused by the division between the arts and sciences in education. The approach of these discussion groups helps to counter this. It's not impossible to overcome: "There's a growing appetite for science — in visitors to the museums for example," said one member. "In my daughter's primary school they try to build on kids' pre-existing notions about the world, like this group does," said another.

Perhaps the most important outcome of these scientific discussions is not so much the gain in technical knowledge but an appreciation of the role of science in the wider political, social and religious concerns that beset the world. Reducing misunderstandings about scientific processes and stereotyping of scientists leaves less room for irrationality and prejudice.

Endnotes

i.   Kuhn, T.S. (2012). *The Structure of Scientific Revolutions*. 4th edition. Chicago, IL: University of Chicago Press.
ii. BBC News (2008). *'Creationism' Biologist Quits Job*, 16 September. Available at: http://news.bbc.co.uk/1/hi/education/7619670.stm.

Chapter 10

Reflections

In the preceding chapters we have considered what makes people want to find out more about science and what possibilities there are for doing so after their school days are over. It has drawn directly on the experience of science discussion groups, whose way of working was described in Chapter 1. In this final chapter, the discussion group experiment is reviewed and some of its broader implications explored.

The Discussion Group Experiment

The idea of setting up discussion groups arose simply from the belief that it ought to be possible for people to talk about scientific ideas in much the same way as they do for ideas in other fields. Surely science doesn't have to be separated off, as a zone for experts alone? The belief that something could be done about this has stuck with me doggedly since childhood. At my excellent primary school in Stevenage, science seemed to me to be perfectly continuous with reading and writing, painting and woodwork — a tribute to my brilliant teachers. One minute I'd be constructing the town church out of matchboxes and lollipop sticks then the next I'd be depicting it in a mural. But as specialisation began to bite at secondary school and training for university exams began to displace the joy of enquiry I felt upset that a choice had to be made between arts and sciences. I watched my friends and fellow students, one by one, dropping science; I began to feel like some kind of survivor in a game of attrition. Those who had dropped science seemed to be just as curious about the world as I was and equally capable of handling complex ideas. It struck me as fundamentally wrong that the subject matter and teaching method were off-putting to so many.

After qualifying in science I decided to move into teaching, to try putting my beliefs into practice. Ultimately, the idea of finding a way to communicate about science beyond the classroom and outside the examination system took hold. The idea of discussion groups — not linked to professional training in science, not constrained by examinations — became a reality, first in formal adult education settings, then in informal surroundings.

As this personal account suggests, a key reason people in England drop science at an early stage has been that it is permissible, sometimes even necessary to do so, at the age of sixteen when the curriculum reduces to just three or four subjects. As a result, most people simply do not sustain their learning in science as long as in other fields. But of course, there are many additional reasons why so few adults feel they have a solid grasp on science: the experience of teaching, the content of the curriculum, the use of language, public and parental attitudes, for example. As we have seen in earlier chapters these reasons commonly crop up in discussion groups when participants reflect on their previous encounters with science; they also emerge in a number of studies about science education. Behind these apparent reasons lie a range of other deeper possible causes: the shortage of science and maths teachers, the white-coat culture of laboratories, syllabuses designed for the university-bound minority and, perhaps, subtle kinds of gender bias.

Despite these negative factors working against the uptake of science in formal education, my belief is that a significant proportion of adults remain fascinated by the world around them and this fascination extends as much to the natural sciences as to the social, political and cultural ones. Infants, as they investigate and theorise about their surroundings with every second that passes, are a huge inspiration. Is it possible that this impulse does not in fact get lost in the process of growing up, but merely suppressed?

It was this possibility that led me to try out an approach to learning about science that capitalised on natural curiosity. The initial intention was, in a way, itself scientific: to spell out a research question, set up an experiment, record what happens and analyse the result. As it turned out the "results" of the experiment were so positive that it developed into an enduring scheme. The key question was: can discussion about scientific ideas, based on issues put forward by ordinary people, lead to better understanding of science? Practical questions followed: if it were tried, would people join in, would it last, would significant learning take place? The whole concept was underpinned by a more profound question: "Is there a way of organising learning in science that mirrors the more person-centred approaches familiar in other subjects?"

To address these questions a method was designed that places the tutor in an intermediary role, standing between the raw questions from everyday life and the vast body of scientific knowledge. The process is described in Chapter 1 and, in summary, involves a sequence of stages in which the tutor:

1. Helps people express and share any negative feelings they hold about their experience of science to date.
2. Encourages people to articulate questions and issues from everyday life that they have always wanted to ask.
3. Facilitates exchanges between members of the groups about the questions, to bring out the perceptions people already have, however well or ill informed.
4. Identifies the key underlying scientific concepts at play in the group discussion, without necessarily answering directly the questions posed.
5. Articulates the underlying concepts (in non-mathematical terms) where possible or logs them to be dealt with later, after further investigation.
6. Facilitates further rounds of discussion about the underlying ideas.
7. Keeps a note of the discussion as it progresses, for circulation later as a reminder for the next occasion.

What the Experiment Shows

The sheer duration of the discussion groups set up along these lines is significant, with participants turning out every month to a discussion about an unpredictable topic with no prescribed outcome. The desire to find out about and understand things seems to be strong and persistent. What is more, new members have joined over time, on the recommendation of existing ones. A few people have left, some have joined, and most have stayed for many years.

The attractiveness of the method to participants demonstrates that it works, at least as a process. People do overcome their hesitations and fears about science; they volunteer questions, reveal their initial understandings and willingly share them. As a group they critique each other's ideas and ultimately develop a strong appetite for scientific explanation. In short, the method develops a strong motivation to engage with scientific ideas. It also demonstrates the possibility of adapting person-centred methods from the arts and humanities in the teaching of science, though it is important to recall that it is only the conceptual aspects of science that are addressed in this approach; other important aspects, particularly the mathematical and experimental ones are missed out. In attempting to extrapolate the learning from this experience more widely across science education, this is an important caveat.

A different learning point from the experiment is how the role of the tutor changes. In this approach the tutor's primary task is to stimulate and sustain self-expression and dialogue between the members of the group rather than deliver a sequence of prepared, authoritative information. In defiance of their training and experience in teaching science, the tutor has to hold back, struggling against an impulse to answer questions directly as they are posed. It can be strange and unnerving no longer being in command of all the knowledge required. By definition, responding to unpredictable questions from everyday life will call for knowledge from any area of the natural sciences (and potentially beyond). Some may be within a tutor's grasp, some only sketchily so, and much will be quite outside their field. This means the tutor must be content with more of a mediating role: mapping out knowledge areas and outlining broad principles in some topic areas, perhaps deferring detailed answers to a later session. The internet makes it possible for members of the group as well as the tutor to look up more detailed information after the session. An example of a shared learning experience is given on the next page.

As it turns out, this altered role of the teacher seems to be understood and accepted by the learners much more readily than a tutor might expect. The idea that questions are not simply to be answered but are to be explored, as in a literary or historical discussion, comes to be understood as a way into deeper understanding. The tutor is constantly moving the discussion towards scientific fundamentals, rescuing discussion from losing its way in contextual details. In the case of quicksand, for example, the challenge was to bring discussion round from images of bad guys drowning in bubbling mud to the chemistry of mixtures and the physics of density. The pedagogic skill is in rapidly identifying the underlying scientific principles at play amongst the welter of particular, and often colourful, everyday experiences.

The most difficult outcome of the experiment to assess is the quality of understanding gained over the longer term. How do we judge whether discursive discussions ranging freely over topics chosen by the learners result in any significant understanding? In the absence of a thorough evaluation of the scheme, the best insight into this comes from focused group discussions with

Sharing the Learning

The nature of quicksand

A member of a discussion group was on holiday on the coast and noticed a warning sign about quicksand. She sent a text message to the group's tutor asking what quicksand was. Not knowing the answer, he looked up the composition and nature of quicksand on the internet in advance of the next group session. The language of the webpages was impenetrable to the group so the tutor's key role was to reinterpret it in understandable terms. Learning about quicksand from various items on the internet became a shared experience between the tutor and the group. Both parties learned together that victims who stray into the substance do not disappear but sink to approximately waist height then float. Someone in the group found a video clip on the spot which showed how a victim can extricate themselves by returning the quicksand to its stable condition. The question stimulated deeper discussion about what it means to float, what is meant by density and how deeply an object, like a ship, sinks in a fluid medium.

participants. Several have offered their thoughts about the longer-term impact of this approach to learning.

My own starting point was to consider what it would be reasonable to expect from a scheme of this kind. Would key principles and facts be remembered or, more broadly, would a sense of the landscape of scientific knowledge be grasped? Would insights into the culture of science and the world of the practising scientist count as valuable outcomes? What about insight into scientific methods? Would an understanding of the nature of scientific knowledge be gained — the idea of models, experiments, theories and multiple levels of explanation?

Here is what a number of participants themselves have said.

I always want to know what is unknown to me. I have noticed others who are content to just accept things (e.g. about electricity) — but when I quiz them they don't really understand them, nor are they bothered about it.

This expresses a clear feature of many discussions: the urge to understand things thoroughly, not just to be told about them. No sooner is one question satisfactorily answered than further, more challenging questions, are asked: "OK, so hormones are chemicals that act on receptors. But how does the body get rid of them after they are used?"

Fully grasping an idea is important for people, rather than merely memorising facts or formulae for an exam. As one participant said:

> When I did voltage for O-level, I did the exam but didn't understand it at all. But discussing our experience of a water tank above a shower — the link between water pressure and height of the tank — made a whole lot more sense to me. Then I could grasp the idea of voltage.

But when topics are based on questions about everyday life and do not follow any kind of syllabus does it add up to anything coherent? Are things remembered? Does a picture emerge or does the scattergun approach just result in fragments of knowledge? Participants are clear that not everything they learn is committed to memory, but they often recall visual images they have acquired, as these comments illustrate:

> At one level I don't remember anything. But I am sure I know more than when I started; for example: electrons moving around and jumping up and down levels in an atom. Previously when science issues arose the shutters came down; now I try to think it through.
>
> We keep coming back to some topics and I remember some things — like free electrons rushing around in a metal. But I still find the concept of charge difficult to grasp.

The responsive, non-systematic approach to choosing topics is generally appreciated:

> I love jumping around, like when we discussed all the various hormones or the elements in the periodic table; in the end it all fits together; life is like that. You always end up with more questions than you ask.

Coherence is described by participants in terms of connections being made rather than a topic being built up systematically:

> Some of what we learn is a lot of bits; some of it gels together. You end up with threads, not layers of understanding — a bit like weaving a textile (with some holes in it). But I am not sure we could explain things to other people. You can just communicate enthusiasm and insights.

Overall, people who have been in discussion groups for many years do not claim to remember much of the detail of what they discussed years ago. After all, how much detail do most of us remember from distant school exams? Some are mildly troubled by this, most are not. What all participants agree on is that it is the process, the quest that really motivates them: the enjoyment of finding out, the satisfaction of grasping something previously baffling. The here-and-now act of learning is the pay-off, not so much a wished-for end result.

Nevertheless, what participants also remark upon is the way in which connections are gradually forged as pieces of knowledge come together. As the emergent curriculum spirals around, key concepts reappear time and again — electrons appear first as components of atoms, then as carriers of electric current, then as bonding agents in a DNA molecule. Each time more of their characteristics are grasped and their existence becomes accepted through familiarity. Through repetition and interconnection, fundamental scientific concepts become part of the mental furniture. However, as participants are eager to point out, understanding reached in this informal way is not so strong that one could explain what one has learned to another person. I expect this is indeed true; but I also suspect that it is not restricted to informal learners of science. The difficulty of trying to explain anything one has just grasped to a third party is immense. One thing I recall clearly from my own university education is the distinguished professor Sir Harrie Massey telling me, a humble representative on the staff–student committee: "If you really want to understand something in physics, try teaching it."

The final point expressed by members of the discussion groups is about the utility of what they learn. In general, they do not see the understanding they gain as being directly applicable; nor does this concern them particularly:

> I come out of interest; I have no opportunity to apply the knowledge I gain; that's no bad thing. We pick and choose topics, we are not systematic; I like that. But you make connections; for example, between how electrical impulses are transmitted along a nerve and the loss of sensation my son had due to an arm injury.

This was an issue about which I had been quite undecided in advance of running the scheme. Would people expect to go away, knowing better how to rewire their house, save on their heating bills or graft their apple trees? As it turns out this has never become an issue in practice. Participants rarely talk about the impact of their newly acquired insights on household practicalities; instead they refer repeatedly to the impact on their confidence to participate in conversations, to follow current affairs, to challenge others parading their superficial knowledge of science.

Scientists and Scientific Culture

Given the opportunity, adults want more from science than just knowledge. As people with experience of the width of the world — of working life, family life, society, politics and economics — as citizens, they want to find out more about the secret world of scientists, how they operate, what makes them tick. They are fascinated by the culture of science itself: the way experiments are organised, the role played by great individuals in the history of science, the limitations of science, the motives of scientists, the lure of fame and fortune. As we saw in Chapter 9, insights into these cultural questions develop seamlessly from discussions that begin with everyday enquiries. Visits to laboratories, viewing of TV documentaries, attendance at lectures and debates — all of these help to acquaint people with the world of the practising scientist, engineer or medical professional, as well as with the specialist knowledge they convey.

The key discovery on entering the world of scientists is that, despite the manner of their portrayal in cinema and on TV, they aren't a different species; in the main, they are rather like you and me. As with other professionals they deal in uncertainties, they have preferences, prejudices and self-interest; on average they are neither saintly nor robotic. Some may find communication difficult at first, when talking about their subject to laypeople, but I doubt if this is significantly worse than with other kinds of expert (in finance or law or post-modern sociology, for example). Experts live in a world of shorthand; the technical language they use eases communications amongst themselves. It just takes a while (and perhaps a few gentle reminders) to shift to a vocabulary and a way of expressing ideas that the layperson can follow. Visits to scientists in the labs and clinics help not only with understanding the topic in hand, but also in experiencing the atmosphere of a scientific workplace and establishing a common language with a scientist. In a visit to a cell biology lab studying the effects of serotonin on depression, participants said it "gave an insight into what real scientists do ... it showed the complexity behind everything". The researcher's explanations were quite understandable (even though they were based on slides from undergraduate lectures) and the sight of depressed nematode worms wriggling around under different doses of serotonin inhibitor gave a real sense of science in action. It also demonstrated that science is not so much "hard" as highly specialised; an expert is highly informed on a particular topic and any outsider feels ill-informed by comparison. Even scientists feel inexpert when straying outside their own specialist field.

As we saw in Chapter 9, not only do scientists and their culture seem less remote when you engage more directly with them, the very nature of science itself becomes more apparent. Perceptions of science as rigid, all-knowing and potentially threatening are gradually modified as insights are gained into *actual* science, as it happens today and how it has developed historically. The limits to what science can be expected to offer begin to be understood as well as the nature of the contribution it makes. Media stories of remarkable medical treatments and extraordinary technology convey an impression of science as all powerful. But closer examination through meeting scientists and exploring scientific ideas reveals it as something humbler. Science is not all-knowing; it doesn't have all the answers. Far from it, the more you learn the clearer it becomes that the realm of the unknown is boundless and appears to grow larger with every discovery. Scientists, like the rest of us, mostly don't know things. What flips the course of the jet stream? What causes depression? Why isn't there a pill for asthma? On the whole, as we probe these questions and gain more insight, further, deeper questions arise. Scientists are constantly toying with uncertainties. A randomly chosen edition of *Scientific American*[i] shows on one page that scientists don't agree on the size of a proton and on another that we can't even model accurately the coiling of a stream of honey dripping from a spoon.

The nineteenth-century sense that science gradually reveals the rigid, knowable laws of nature has given way to a more utilitarian view. Science is seen to derive from experimentation and observations, from which provisional generalisations can be made and models devised to explain the observations. The nature of light, for centuries the subject of rivalry between particulate and wave-like theorists, is now understood pragmatically as a matter of choice. The particulate or photon model is used to explain some observations, the electromagnetic wave theory to explain others. Models explain the observations we have made to date and are useful insofar as they make predictions that are subsequently confirmed. When and if they cease to do so, they are replaced. This is how Einstein replaced Newton in explaining gravitation, nucleic acids replaced proteins as the home of genes, and tectonic plates replaced volcanic activity to explain the movement of continents. The realisation that scientific phenomena can be explained through models as well as laws has proved enormously liberating in discussion groups. As one participant expressed it "realising there were models, not just laws and facts, was one of the eureka moments in the ten years of our discussions". Now the metaphor for imagining electrons in an atom could be either "shells" or "clouds" or "energy levels", according to the question being asked. By accepting the ambiguity in scientific understanding many obstacles are removed. As a parent in one group once remarked:

> Couldn't teachers be more explicit about the existence of different models? When my daughter was given different explanations in biology as she moved from lower to upper school, she said: "So they were lying to me before."

So, although the exploration of scientific ideas in an informal, reality-based way is unlikely to lead to a comprehensive grasp of the scientific disciplines, it does enable people to enter into the spirit of science, to grasp something of its methods, its culture, its history as well as some of its more fundamental concepts. What is more, it does this in a manner that enthuses and motivates where more formalised approaches have failed to do so.

Science as a Subject

So where does the experience of informal science discussion groups leave the major question of science as a subject? Is there a way of re-conceptualising the way it is represented in education? Could something of the sense of excitement and discovery that characterises actual science be more widely communicated?

The problem doesn't lie in the nature of science itself but in the way it comes across. As we saw in Chapter 3, it tends to be associated at both school and university with professional preparation, in a way that history or geography, English or philosophy are not. Science education is organised to serve the scientific career. The consequence is that curricula and examinations tend to be too technical to appeal to the majority of students. Unsurprisingly, far too many people drop the subject as soon as the opportunity arises; most feel excluded from the area and, in effect, they are. The result of so many failing to learn about basic scientific ideas in the early teenage years, and then ceasing to try thereafter, is an adult population with a weak grasp of the main concepts. In a democratic society in which political choices frequently include a scientific aspect, this is worrying.

Is it conceivable therefore that a different kind of school subject might be developed based around scientific knowledge and method, not designed to meet the vocational needs of the professions? Could an analogy be drawn with history or geography, sociology, religious education (RE) or music — a school subject that draws on life experiences and prepares in a general way for adult life? Could the vocational training of future scientists, engineers and medical professionals be organised separately — perhaps as a parallel stream, or maybe deferred for later specialist study, as for law or architecture?

Such ideas about a fundamental revision of science at school are certainly not new. A major report for the Nuffield Foundation[ii] on science education across the EU concluded that:

> Science education for all can only be justified if it offers something of universal value for all rather than the minority who will become future scientists. For these reasons, the goal of science education must be, first and foremost, to offer an education that develops students' understanding both of the canon of scientific knowledge and of how science functions. In short that school science offers an education in science and not a form of pre-professional training.

Indeed, the Nuffield Foundation has worked with a group of organisations to develop a suite of GCSE syllabuses, Twenty-First Century Science, building on these ideas. In a similar vein, a report from a major research programme in the UK[iii] recommended that:

> The provision of a high quality "science education for citizenship" for all students should continue to be energetically addressed. Its establishment would, we hope, help more students to see the intrinsic worth of a career in science-related fields.

Given such authoritative calls for a new, more open kind of science curriculum at school level, does the experience of these adult learner groups, which do offer "something of universal value", have something to offer? Which insights from the adult experience could be of more general use?

Pedagogy

Perhaps the clearest message from the experiment is the power of everyday questions as the starting point. In a suitably encouraging atmosphere, people (or at least adults) are shown to be ready and able to articulate questions based on everyday experience, which, given a suitably equipped teacher, can inspire learning about fundamental scientific concepts. Links to everyday experiences do much more than simply lend colour to dry theoretical explanation; they can act as the point of departure and as reference points as scientific concepts are explored.

The experience also demonstrates a further, equally important pedagogic point — that opportunities for self-expression and learning from one's peers can be created in science as in other subjects; transmission teaching and learning by rote are not the only way. Dialogue between the intuitive theories that we all carry with us and concepts that have been established scientifically not only motivates people to engage with a topic but also tends to deepen understanding.

The success of the scheme also suggests that a form of science learning is possible without the use of mathematics or laboratories; the emphasis on ideas and critical thinking rather than mathematical deduction or practical experimentation works. Having said this, the absence of a practical aspect to the scheme is purely a consequence of the lack of laboratory facilities for the general public. The opportunity to manipulate equipment and try out practical procedures would greatly enhance adult learning experiences, even when the main emphasis is on conceptual understanding.

Beyond these general observations about the scheme, what can be usefully said about the specific pedagogic approaches adopted in the freer context of adult learning? The most important of these is the attention given to people's pre-existing conceptualisations. Where these contain errors or misunderstandings it is common practice to seek to simply replace them through instruction. In this approach, by contrast, they are treated respectfully as offering clues to where conceptual problems lie; they provide diagnostic information. Two examples in which people freely reveal misunderstandings illustrate the point:

> The cold weather was being discussed which led into the science of freezing. When the freezing point of alcohol was contrasted with that of water, one participant said: "It's difficult thinking of something freezing not at zero degrees."
>
> In discussion about the nature of hormones, a person volunteered: "Now we know that steroids are simply a chemical group it changes your view of them as a bad thing because of doping."

The need to listen carefully to how people express themselves calls for considerable restraint on the part of the tutor, who must avoid pitching in with the scientific answer too early. An important observation from the experience of discussion groups is just how much progress can be made towards the scientific concept through peers discussing the strengths and weaknesses of their own preconceptions. People in the group may be better able than the tutor to empathise with their colleagues' perceptions. By attempting to critique each other's ideas they may also get to see more deeply into the concept. The challenge for the tutor is to hold back from authoritative explanation until the need for it is overwhelming.

A further pedagogic challenge is to avoid answering specific questions too directly. If discussion is reduced too often to brief exchanges — "Which is the live wire in an electric plug?" or "What is a stem cell?" — the opportunity to learn about more generalisable concepts is lost. A particular task for the tutor is to spot the underlying scientific principle behind a question and to bring this out before applying it to the question in hand.

Curriculum

In addition to these conclusions about the teaching approach, what can be said about a curriculum that emerges in retrospect rather than being laid down in advance?

Although in the short term topics are selected in unpredictable ways, over time a curriculum emerges that spirals around topics rather than marching steadily through them. Starting points differ markedly from session to session, reflecting the plurality of everyday life, but the scientific fundamentals they lead to are relatively few. They get addressed repeatedly in the varying contexts of the questions posed. Ultimately, the fundamental concepts are treated in much the same way as they would be in any teaching environment. Information about atomic structure, chemical bonds, DNA transcription, nerve transmission is presented in broadly conventional ways. What differs, however, is the tendency for people to want to challenge the knowledge as it develops: How do we know this? Why is this the case? If this is the case, what about that? This more mature approach to learning means that aspects of a concept that might otherwise baffle the learner tend to get probed relentlessly.

This heightened awareness of the learning process frequently extends to questioning the nature of scientific knowledge itself. As we saw in Chapter 9 many aspects of the scientific culture, its methods and history, spin off from discussion about particular concepts. Thus, a session on so-called "junk DNA" led to questions about how scientists deal with the unknown; the development of tectonic plate theory showed the value of different disciplines working together; a session on the transformative effect of digital technology led to a historical review of the role of emerging technologies in scientific theory.

The curriculum that emerges turns out to be far less fragmented than one may at first suppose. By focusing on underlying concepts, reached through specific contexts, the range of issues is relatively constrained — the structure of matter, nature of the universe, mechanisms of communication, evolution and the action of nerves, hormones, cells and biological molecules form the major areas of interest. The contexts in which these fundamentals are discussed do, however, vary enormously — from musical performance to child development, digital technology to pharmacology.

Summarising the various outcomes of this experimental approach, in relation to scientific content, style of teaching and impact on learners, we can say that it demonstrates how adults can learn about scientific ideas in ways that engage them more as a person when suitable approaches are adopted.

Wider Implications

The overall approach used in this experimental scheme differs radically from that of most formal science education, above all in its lack of a pre-determined curriculum. However, many aspects of the approach are not new or untested. Links to everyday life, encouragement of peer-group discussion and expression of prior understanding, for example, are practices in common use in effective teaching today and have been the subject of much educational research and innovation. Indeed, many initiatives have been organised over the decades to reform science syllabuses and alter approaches to teaching. Some have focused on linking syllabuses more closely to real life, others on self-directed learning; some have helped teachers with more imaginative ways of teaching and others have encouraged women to take up science. A recent review of research on teaching and learning for the UK's Institute of Physics[iv] concludes:

> Studies that have examined pedagogic strategies that increase students' motivation and enjoyment to learn science have found that students give value to approaches that are typical of teaching and learning in a humanistic science curriculum. The strategies that maintain students' autonomy and responsibility for their learning include investigative laboratory work, group and class discussions where alternative views are considered and valued, problem-solving and project-based activities where students are the decision makers. … There is emerging evidence that these strategies impact positively on the achievement levels of girls as a group relative to boys and have no negative impact on boys' overall achievement.

There is a widespread wish to increase the number of young people, particularly women, studying science subjects; indeed, it is a primary issue for the many organisations concerned about the future workforce in science, engineering and medicine. The growing "public engagement" movement is also helping to raise public awareness of the scientific background to issues such as climate change, gene therapy and energy supply. Although learning may not be the primary aim of such engagement activities, many of the same techniques are in play — starting from people's real concerns, facilitating open discussion and communicating scientific concepts in ordinary language.

The case for a new kind of science curriculum that genuinely appeals to all, rather than the few who will go on to science-based careers, has been made repeatedly by science education experts. However, so far it has not proved possible to make this happen on a large scale. The experience of the small scheme that has inspired this book illustrates some aspects of a more open approach and provides encouragement for those looking for a new way. But to reform the way science is handled within the education system as a whole would be a major national challenge, one that has been set out clearly in a major review in Australia's *Reimagining Science Education*.[v] To bring about actual change an unprecedented degree of consensus would be required about the value of a broadly based subject for all, in place of the existing preparation for science-based careers. This would be difficult enough to achieve amongst education professionals, universities and professional bodies; it would be even harder between the political parties. Even if consensus about the principle were to be established, many practical issues would have to be addressed. New curricula would have to be specified with the agreement of government, examining bodies, universities, employers and scientific institutions. Modified arrangements would have to be made in parallel for the minority who proceed to science-based careers. The existing teaching force would need in-service training in the new approach and an expanded workforce of new teachers would need to be recruited for the more broadly based subject. New textbooks and learning materials would need to be devised and produced to support the new subject — a considerable investment by publishers and the public purse alike. Despite these practical and political difficulties the task is not impossible, as Twenty-First Century Science[vi] has demonstrated by offering one course that develops the scientific literacy of citizens and another that, taken together with the first, prepares young people for more advanced studies in science, should they choose it. Learning materials and support for teachers have been developed alongside the syllabus specifications.

Change can also occur outside the mainstream school system. Adult education, public engagement activities and enhancement schemes for schools provide opportunities to introduce more open-ended approaches designed to attract greater numbers of learners to science. If these areas were to grow in importance the impact, though gradual, could nevertheless be of great importance to society at large, leading to more widespread understanding of scientific concepts. But, as the experience of this scheme shows, the impact would go much further than this. The real world of uncertainty and probability would be more widely appreciated in contrast to the fixed ideas and dogma that are so common. The importance of reasoning, of working up hypotheses and subjecting them to test, of explaining the world through models rather than prescriptions would be more widely acknowledged and perhaps understood. People would become familiar with the methods used by scientists and how scientific knowledge is acquired; the world of science would appear less intimidating, as the scope for misunderstanding and prejudice diminished. With such changes to the shape of science as ordinary people experience it, perhaps the subject might even become as attractive to girls and women as it is to boys and men; this change alone could transform the crisis many feel is looming over the supply of technically and medically trained people for the future workforce. Even more significantly for society in general, the scientific aspects of the political and social problems the world faces might be embraced as vigorously as all the other aspects.

Endnotes

i.   Editorial (2014). *Scientific American*, 310(2), 1.
ii. Osborne, J. and Dillon, J. (2008). 'Science education in Europe: critical reflections. A report to the Nuffield Foundation.' Nuffield Foundation, London. Available at: http://www.nuffieldfoundation.org/science-education-europe.
iii. TLRP (2011). *Science Education in Schools: Issues, Evidence and Proposals. A Commentary by the Teaching and Learning Research Programme.* Available at: www.tlrp.org/pub/documents/TLRP_Science_Commentary_FINAL.pdf.
iv. Murphy, P. and Whitelegg, E. (2006). *Girls in the Physics Classroom: A Review of Research of Participation of Girls in Physics.* London: Institute of Physics.
v.   Tytler, R. (2007). *Re-imagining Science Education*, p. 64. Available at: http://research.acer.edu.au/aer/3.
vi. For details see: http://www.twentyfirstcenturyscience.org/?q=about/about-twenty-first-century-science.

Index